社区市场建筑
Community Markets

[丹] BIG建筑事务所等 | 编

于风军 曲洪波 侯逸宁 曹艺馨 周淼渺 辛敏裕 冯源 周美含 等 | 译

大连理工大学出版社

社区市场建筑

004　自然生长式建筑设计 _ Ljubomir Jankovic

新项目

008　乐成四合院幼儿园 _ MAD Architects
026　凤凰中央公园 _ Durbach Block Jaggers Architects + John Wardle Architects
042　梅罗佩学校 _ XDGA
056　福特科技园 _ RANDJA - Farid Azib Architects

超级自然景观，为新世代而生

070　超级自然景观，为新世代而生 _ Phil Roberts
076　Lune de Sang展馆 _ CHROFI
100　山海美术馆 _ gad
118　周口店北京猿人洞1号考古遗址保护棚 _ THAD
136　艺术生物栖息地水景园 _ Junya.Ishigami + Associates

社区市场：文明的纽带

148　社区市场：文明的纽带 _ Herbert Wright
154　阿泽尔格河畔拉米尔的有顶市场 _ Elisabeth Polzella Architecte
166　格拉莫罗特市场广场 _ Niro Arquitectura + OAU
176　圣雷莫安诺纳里奥市场重建 _ Calvi Ceschia Viganò Architetti Associati
186　胜利市场临时安置点 _ LUO Studio
196　纳吉克罗斯市场大厅 _ Kiss-Járomi Architect Studio
208　特拉斯科工匠市场 _ Vrtical

220　建筑师索引

004 Architecture of Biologically Grown Buildings _ Ljubomir Jankovic

New Projects

008 YueCheng Courtyard Kindergarten _ MAD Architects

026 Phoenix Central Park _ Durbach Block Jaggers Architects + John Wardle Architects

042 Melopee School _ XDGA

056 FORT Technology Park _ RANDJA - Farid Azib Architects

Hypernatural Landscapes for Future Generations

070 Hypernatural Landscapes for Future Generations _ Phil Roberts

076 Lune de Sang _ CHROFI

100 Mountain & Sea Art Museum _ gad

118 The Protective Shelter of Locality 1 Archaeological Site of Zhoukoudian Peking Man Cave _ THAD

136 Art Biotop Water Garden _ Junya.Ishigami + Associates

Community Markets: Links in Civilization

148 Community Markets: Links in Civilization _ Herbert Wright

154 The Covered Market of Lamure-sur-Azergues _ Elisabeth Polzella Architecte

166 Gramalote Market Square _ Niro Arquitectura + OAU

176 Restructuring of the Sanremo Annonario Market _ Calvi Ceschia Viganò Architetti Associati

186 Temporary Site of Shengli Market _ LUO Studio

196 Market Hall of Nagykőrös _ Kiss-Járomi Architect Studio

208 Tlaxco Artisans Market _ Vrtical

220 Index

自然生长式建筑设计
Architecture of Biologically Grown Buildings

Ljubomir Jankovic

　　目前，全球面临着气候变化和自然资源滥用问题。但即便如此，建筑设计和翻新仍采用传统方式，模式流程守旧低效。在全球二氧化碳总排放量中，建筑的排放量就占到了30%。通常，建筑师会提出几个备选设计方案，并形成最优设计以供最终开发。但这种设计方式会最大限度地限制对广大设计可能性的探索，并常常逆气候而动，最终得到的只是次优性能。

　　但是我们可以想象这样一个世界吗？建筑设计从建模开始，逐步探索设计空间，最终形成设计成果，恰如自然生物有机体的生长一样。建筑设计应如何复制生物发育过程，与气候协调一致，将建筑对环境的影响降至最低，完全改变目前的设计理论和实践文化？

　　在人类定居史早期，人们就地取用建材，构筑宜居外形，搭建容身之所。人们取用动物皮革和木棍搭建窝棚，伐修树木、雕刻岩石，或捏泥成形、铺草固定，这些都是人类建造庇护所所需的材料。整个过程基本是自上而下的，即人们首先对要建造的建筑设置一个最终愿景，然后再寻找材料实现这个愿景。然而在接下来的一千年中，技术和科学的进步并没有改变这种自上而下的思维模式，它深深植根于我们的潜意识中。发展自然科学的目的是用一般规律来解释我们周围的世界，而发展传统数学的目的则是将各方面的系统归结为一个整体，并加以表达。甚至到计算机时代初期，这种做法仍然根深蒂固：人们继续将系统作为一个整体，积极发挥计算机模拟技术快速处理模型的优势，自上而下地开发模型。同样地，

Despite the current challenges of climate change and overuse of natural resources, with buildings contributing to 30% of global carbon dioxide emissions, buildings are still designed or retrofitted using old and inefficient processes in a "business as usual" way. Typically, several candidate designs are created by an architect, and the most favorable design is put forward for final development. This leads to a minimum exploration of the vast space of design possibilities, frequently working against the climate and achieving sub-optimum performance.

But can we imagine a world in which building designs are grown into ultimate configurations from design embryos, like nature grows biological organisms, exploring the design space gradually and thoroughly? How can we replicate the processes from developmental biology, working with the climate, minimizing the environmental impact, and completely changing the culture of the current design theory and practice?

In the early days of human habitation, buildings were created from available local materials, shaped into suitable forms. Animal skins and sticks were used to create shelter; trees were cut and trimmed and rocks were carved or earth was shaped and enforced with straw to create materials suitable for making enclosures for human habitation. The process was essentially top-down, with an ultimate vision of the building to be built and with sourcing the materials to fulfill that vision. Advances in technology and science over forthcoming millennia did not change the top-down mindset, which was deeply rooted in our subconsciousness: natural sciences were developed to explain the world around us in terms of general laws and traditional mathematics was developed to represent systems as a whole. This approach persisted even at the onset of the computer age: computers were used to continue developing top-down models of systems as a whole, with an advantage of rapid processing of the models through simulation. The same top-down approach has been used in building design, resulting in a superficial exploration of the vast design space.

This top-down approach is in sharp contrast with how things work in nature. Biological systems develop from the bottom-up: stem cells of an embryo divide and grow in their number and size, differentiating from their original function and specializing for functions suitable for fulfilling

建筑设计也运用了这种自上而下的方法,导致业界对广泛设计空间的探索仍停留在表面上。

这种自上而下的方法与自然界的运作方式形成了鲜明对比。生物系统是自下而上发展的:胚胎干细胞分裂,不断增多,逐渐变大,异化原有功能,形成专用于完成特定任务的功能,如肌肉收缩、神经元信号传输等。随着细胞的进一步生长,它们按相似度形成组织,组织再形成器官,器官最终形成有机体。

上述复杂的系统内自组织行为来自于多个单元的交互作用,且每个单元按照简单的规则运行。这种自组织行为被称为"涌现"。鸟群等多单元系统可以有多种配置。以约翰·C.雷诺兹[1]发现的简单鸟群为例:该鸟群由37只鸟组成,彼此之间有所交互,那么按照我的计算,其设计空间的数量级,可与大爆炸以来的秒数数量级相当。[2]因此,一个简单的鸟群便极其复杂,使得我们无法通过传统数学为其搭建自上而下的系统模型。然而,当我们按照单元交互等简单规则,为鸟群搭建自下而上的模型时,却能发现这种模型在笔记本电脑上就可以无缝运行,并自组织成群体式结构。[1]

自上而下和自下而上都有一个共同点,即生成的系统由各部分的总和组成。关键的区别在于,自下而上的开发是逐步探索设计空间的可能性,并在过程的每一步都以最高的能源资源效率进行开发,而自上而下的开发则是有限的设计空间探索。

生命有机体和数学过程贯穿着一条共同主线,那就是数学概念。[3]最初艾伦·图灵[4]研究的化学物质

specific tasks, such as contraction of muscles, signal transmission in neurons, and others. As they grow, they create groups of similar cells that form organs, and the groups of organs form organisms. This complex self-organized behavior of a system that arises from the interaction of a number of constituent components where each component is governed by simple rules is named "Emergence". A system consisting of several constituent parts, such as a flock of birds, can take a number of configurations. My calculations show that a simple flock consisting of 37 birds and interacting with each other as discovered by John C. Reynolds[1] has the design space that is commensurable with the number of seconds since the Big Bang[2]. Thus, a simple flock of birds has exceptionally high complexity, and a top-down model of such system based on traditional mathematics has not been built. Yet, a bottom-up model of a bird flock based on simple rules of component interaction runs seamlessly on a laptop computer and self-organizes into a flock-looking structure.[1]

Both top-down and bottom-up approaches have one aspect in common, in that the resultant systems consist of the sum of parts. The key difference is that bottom-up development explores the space of design possibilities gradually, taking the most energy and resource efficient state at every step of the process, whilst the top-down approach has a limited scope for design space exploration.

Mathematical concepts have been seen as a common thread between living organisms and mathematical processes.[3] Periodic patterns of concentration of chemicals, first studied by Alan Turing[4], are now understood to be the primary drivers of the morphological development of embryos.[5] John von Neumann[6] developed models of biologically-inspired self-reproducing machines named Cellular Automata, subsequently used to model shell shapes and animal pigmentation patterns. Stephen Wolfram[7] devised the Principle of Computational Equivalence stating that: "all processes, whether they are produced by human effort or occur spontaneously in nature, can be viewed as computations". Thus, "no system can ever carry out explicit computations that are more sophisticated than those carried out by systems like Cellular Automata and Turing

浓度周期性模式现在被认为是胚胎形态发育的主要驱动因素。[5]约翰·冯·诺依曼[6]受生物学启发，构思了一种自复制机器 (Cellular Automata, 元胞自动机)，随后该构思被用于建模外壳形状和色素沉着动物模型。史蒂芬·沃尔夫勒姆[7]则设计了计算等价性原理："所有进程，无论是人为推动的还是自然发生的，都可以被看作是计算"，因此"没有任何系统能执行比元胞自动机和图灵机等系统更复杂的显式计算"。由于元胞自动机/图灵机是基于简单规则的，因此胚胎的发展可以解释为基于简单规则的计算。

人们通过仿生学等一系列组合措施，试图在建筑设计中落实生物学带来的灵感。仿生学研究所[8]对仿生学的定义是："仿生学是一种创新方法，通过模仿大自然长久运行的模式和策略，寻求可持续的解决方案，应对人类面临的问题挑战。"迈克尔·波林[9]在艾登项目 (2015年)[10]中通过仿生技术，在形式、功能和资源方面实现了高效益成果。其中，项目六边形结构的灵感就来自肥皂泡、花粉粒和碳分子形状。这解决了在不规则场地上建造大型温室穹顶的难题，提高了六角形材料的使用效率，进而提供了更大的太阳能光圈，增加了太阳得热。然而，该项目和其他受自然启发的建筑，如Hensel等设计的建筑[11]，并没有真正复制生物进程。

这就逐渐引出了本文问题的答案。建筑由许多部分组成，并且具有无限的设计空间。但是一直以来所使用的自上而下的设计会限制设计空间探索，导致次优设计。不过，当前的自上而下式建筑设计仍然能力不足，除了仿生学等松散的生物学关联，无法在单元层次等不同组织层次上实现适应性强的自组织，更无法实现不断增长的设计。这需要系统地改变设计思维方向，还要改进设计工具，推动这种方向变化。我

Machines". As Cellular Automata/Turing Machines are based on simple rules, it follows that embryonic development could be interpreted as a computation based on simple rules.
Attempts towards biological inspirations in architecture have been made by a combination of approaches called biomimicry. As defined by the Biomimicry Institute[8], "biomimicry is an approach to innovation that seeks sustainable solutions to human challenges by emulating nature's time-tested patterns and strategies". Michael Pawlyn[9] used biomimicry for achieving form, function and resource-efficiency in Eden Project (2015)[10], where soap bubbles and the shape of pollen grains and carbon molecules inspired the hexagonal structure of the project biomes. That solved the problem of creating a large greenhouse dome on an irregular site, leading to material efficiency of the hexagonal shape that provided a greater solar aperture and increased solar gain. However, this and other instances of nature-inspired architectural forms, such as by Hensel et al.,[11] were not a true replication of biological processes.
Thus, the answers to the questions raised in this article are starting to emerge. Buildings consist of numerous parts, and have hyper-astronomic size of design space, but they are always designed from the top-down, resulting in a minimum exploration of the design space and consequent sub-optimum designs. The current top-down building design thinking, despite the occasional loose biological associations such as biomimicry, simply does not have the capacity for adaptive self-organization at a variety of organizational levels on a component level to create growing designs. This requires a systemic change of direction in the design thinking and a change in design tools to facilitate that change of direction. These changes will come only if significant research efforts are invested into the study on interpreting the rules of developmental biology and replicating them within the framework for building design that makes them available to architects and helps them to change their ways of designing. It is only then when the simple rules hidden in the depth of the self-organization of the natural world around us will give rise to designs that explore the design space gradually and thoroughly.
Hence, architects would swap their drawing board or a CAD system for a new biologically-aided computer design framework, where they would plant embryonic building cell seeds, set the external and internal environment conditions, set the stopping criteria for growth such as the floor area, and let the building design grow. Through cell to cell interaction, where the cells will

们要投入大量精力进行研究，解释发育生物学规则，并在建筑设计框架内复制规则，让它们有用于建筑师，推动他们改变设计方式。只有这样，变化才有可能出现。我们要深入挖掘周围自然世界隐藏的简单组织规则，也只有这样，才能逐步、全面地探索设计空间。

由此，建筑设计师将图纸或CAD系统更新为全新的生物辅助计算机设计框架。在此之上，他们将"种植""建筑细胞胚种"，设置内外部环境条件，设定建筑面积等增长限停条件，再放手让建筑设计不断增长壮大。通过彼此间的相互作用，建筑细胞将找到最合适的位置和功能，传递并吸收能量，或防止能源损失，配合内外部条件和建筑中的定位，形成最终的设计。建筑师可以重复这个过程，改变初始条件和胚种DNA，直到实现满意设计。

这种新的设计思维（即建筑设计按设计细胞逐渐生长壮大）将最有效地利用能源，形成有效的配置，形成创新和富有想象力的解决方案，满足我们的居住需求，保护我们的星球不受资源过度使用和气候变化的影响。在生物学框架下进行的建筑建造将运用机器人技术和3D打印技术。这将是新的开端，是一场智力和技术革命的开端，也是我们文明发展的重大变革。我们要以开放的心态拥抱新的方式，并将旧的方式融入其中，这样我们对建筑的理解、欣赏水平将大大提高，使它们更接近生物存在的形式。

我提出的是一种不太可能实现的新型理想模式吗？我不这么认为。考虑到生物系统已经存在超过46亿年了，没有什么终极障碍可以阻止我们以生物方式改进建筑设计。

be adopting the most appropriate position and function, specializing in transmitting energy, absorbing energy, or preventing energy loss, in response to internal and external conditions and their position in the building, the ultimate design will emerge. Architects could then repeat the process, changing the initial conditions and embryonic cells "DNA", until a satisfactory design is grown.

Under this new design thinking, as building designs grow a design cell by a design cell, they will take the most energy and shape efficient configuration at every step, creating innovative and imaginative solutions for fulfilling our habitation needs and for protecting our planet from over use of resources and climate change. Once they are grown in a developmental biology inspired framework, such buildings will be constructed using robots and 3D printers. This will be an onset of a new paradigm that could be considered as a start of a new intellectual and technological revolution, and a step change in the development of our civilization. As our mindset embraces the new ways and puts the old ways into perspective, our understanding and appreciation of buildings will increase, making them resonate closer with our biological being.

Am I proposing a new type of utopia, something that is unlikely to happen? I don't think I am. Considering that biological systems have been doing this over 4.6 billion years, there is no ultimate barrier that would prevent us from improving building design process in a similar way.

1. Reynolds, C. (1987). Flocks, Herds, and Schools: A Distributed Behavioral Model. *Computer Graphics*, 21(4), 24–34. http://www.cs.toronto.edu/~dt/siggraph97-course/cwr87
2. Jankovic, L. (2012). An Emergence-based Approach to Designing. *The Design Journal*, 15(3), 325–346. https://doi.org/10.2752/175630612X13330186684150
3. Sharpe, J. (2017). Computer modeling in developmental biology: growing today, essential tomorrow. *Development*, 144(23), 4214–4225. https://doi.org/10.1242/dev.151274
4. Turing, A. M. (1952). The chemical basis of morphogenesis. *Philosophical Transactions of the Royal Society of London. Series B, Biological Sciences*, 237(641), 37–72. https://doi.org/10.1098/rstb.1952.0012
5. Kunche, S., Yan, H., Calof, A. L., Lowengrub, J. S., & Lander, A. D. (2016). Feedback, Lineages and Self-Organizing Morphogenesis. *PLOS Computational Biology*, 12(3), 1–34. https://doi.org/10.1371/journal.pcbi.1004814, https://plato.stanford.edu/archives/fall2015/entries/biology-developmental
6. von Neumann, J. (1967). Theory of Self-Reproducing Automata. John von Neumann. Edited by Arthur W. Burks. University of Illinois Press, Urbana, 1966. p.408, illus. $10. *Science*, 157(3785), pp.180–180. https://doi.org/10.1126/science.157.3785.180
7. Wolfram, S. (2002). *A New Kind of Science*. Wolfram Media. https://www.wolframscience.com
8. Biomimicry Institute. (2006). *What Is Biomimicry?*. https://biomimicry.org/what-is-biomimicry
9. Pawlyn, M. (2010). *Using nature's genius in architecture*. https://www.ted.com/talks/michael_pawlyn_using_nature_s_genius_in_architecture
10. Eden Project. (2015, February 12). *Eden story*. https://www.edenproject.com/eden-story
11. Hensel, M., Menges, A., Weinstock, M. (2013). *Emergent Technologies and Design: Towards a Biological Paradigm for Architecture* (1st ed.). Routledge. https://doi.org/10.4324/9781315881294

乐成四合院幼儿园
YueCheng Courtyard Kindergarten

MAD Architects

幼儿园体育场屋顶"漂浮灵动",将传统北京四合院"捧在手心"
A floating kindergarten playground roof surrounds a traditional Beijing courtyard building

四合院属于北京传统建筑,整体布局围绕院落展开。乐成四合院幼儿园所处的场地上,有乐成教育集团使用的一座四合院(公元1725年建成),一座与其相邻的建于近期的四合院,以及一栋20世纪90年代四层L形建筑。幼儿园由当地MAD建筑事务所设计,占地8000m²,于2019年开放使用。项目设计由事务所的负责人马岩松领导,他就出生于北京。在他的领导下,设计团队在场地中插入了一座单层建筑。它采用灵动的浮动屋顶,将历史悠久的四合院建筑"捧在手心",不仅保护了场地上的文化遗产,还讲述了新旧共存的多层次城市叙事故事。

幼儿园屋顶柔和起伏,延伸面积近3500m²,高度与四合院的老式山墙屋顶大致持平,并将不同建筑之间有限的空间转变为操场,供400名2~5岁的儿童游乐玩耍。它采用了不同色调的红色,并逐渐过渡到白色。在屋顶西南区域,起伏不定的高度还形成了几个小小的"山丘"和"平原"。在屋顶的北侧,最高的"山丘"形成了一个像弯顶一样的结构,孩子们可以从一侧进入其中。屋顶上的五处弯曲洞口在下面形成了新的玻璃小院,将自然光引入室内。这些小院使得三棵老槐树被保留下来,老槐树与四合院内的树木一起,给孩子们以树端漫步的感觉,引领孩子们亲近自然。其中一处洞口处设计了弯曲楼梯和扭曲滑管,孩子们可借此从屋顶上滑下来。屋顶边缘有栏杆保护孩子,屋顶表面采用了类似橡胶的塑料,多孔防潮。

"漂浮"的屋顶之下,是幼儿园设有落地玻璃窗的开放式布局空间。开放空间形成了自由和包容的氛围,也承载了日常教育的功能。这一新的学习空间紧邻老四合院,面向历史建筑开放,为孩子们提供了新旧交替的视野,加深了他们对时空的理解。北侧地下另有一座环形封闭小剧场和一座体育馆。四合院建筑内部被重新利用,设有教职员工设施和艺术舞蹈教室。

MAD建筑事务所的设计真实地保留了原有建筑,包括高层楼房。马岩松认为,这座高层建筑正是20世纪90年代的"历史印记"。它被设置为备用教室和办公室。新建筑与各早期建筑之间形成了新的互动,增加了社区的价值。同时,孩子们将加深对环境、历史和本地的理解,提高环保意识。

A siheyuan is a traditional Beijing building which encloses a courtyard. The YueCheng Courtyard Kindergarten is on a site where an education group occupies a siheyuan which dates from 1725, an adjoining siheyuan built recently, and a four-story L-shaped block from the 1990s. The new 8000m² structure designed by locally-based MAD Architects opened in 2019. The design, led by the practice's Beijing-born principal Ma Yansong, inserts a one-story building into the site, featuring a dynamic floating roof that surrounds the historic courtyard building. While preserving the cultural heritage of the site, it forms a multi-layered urban narrative, where old and new co-exist. The softly undulating roof extends over almost 3500m², approximately level with the old gabled roofs of the siheyuan, and transforms the limited space between the various buildings into a playground for 400 children aged 2-5. It is colored in different shades of red transitioning to white. The roof's fluid terrain has several small hills and plains on its south-west area. On the north side, the highest hill creates a dome-like pavilion which can be entered on one side. Five curved openings create small new glazed courtyards below, bringing natural light into the interior. These courtyards allow three old pagoda trees to be retained. Along with the siheyuan trees, they bring the children proximity to nature, like a tree-walk. One opening hosts a curving staircase beside a twisting tube which allows children to slide down from the roof. The roof edges have railings to protect the children, and the roof surface is a rubber-like plastic which is porous to moisture.

Below the floating roof, the kindergarten is an open-concept layout with full-height glazing. The openness of the space offers a free and inclusive atmosphere, and functions as the daily education space. Positioned adjacent to the old courtyard, the new learning space opens towards the historic buildings. This gives the children alternating views between old and new, deepening their understanding of time and dimension. A small enclosed amphitheater in a circular space and a gymnasium are sunk below ground level on the north side. The siheyuan building interiors are repurposed to include staff facilities, and classrooms for art and dance.

MAD's design retains the authenticity of the original buildings, including the tall block, which Yansong felt was a "history stamp" from the 1990s. It is designated for additional classrooms and offices. The new structure creates a new kind of interaction between all the earlier buildings and adds value to the community. The children will gain an understanding of the environment, of history and place, and the preservation of nature.

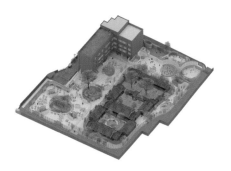

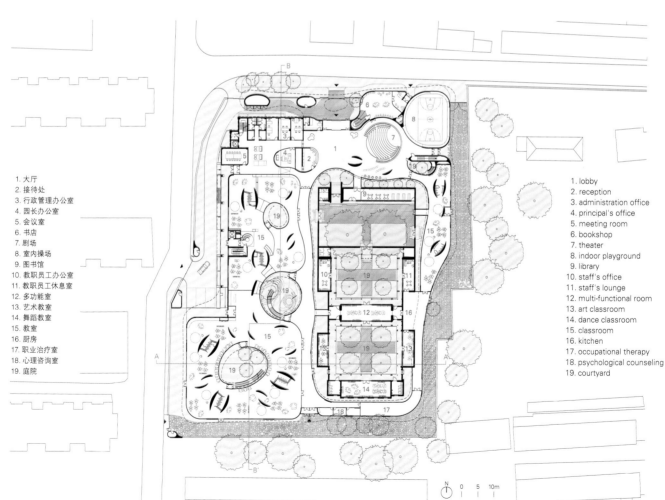

1. 大厅
2. 接待处
3. 行政管理办公室
4. 园长办公室
5. 会议室
6. 书店
7. 剧场
8. 室内操场
9. 图书馆
10. 教职员工办公室
11. 教职员工休息室
12. 多功能室
13. 艺术教室
14. 舞蹈教室
15. 教室
16. 厨房
17. 职业治疗室
18. 心理咨询室
19. 庭院

1. lobby
2. reception
3. administration office
4. principal's office
5. meeting room
6. bookshop
7. theater
8. indoor playground
9. library
10. staff's office
11. staff's lounge
12. multi-functional room
13. art classroom
14. dance classroom
15. classroom
16. kitchen
17. occupational therapy
18. psychological counseling
19. courtyard

赫伯特·怀特对马岩松的采访
Interview_Ma Yansong + Herbert Wright

赫伯特·怀特（以下简称赫）：当提出乐成四合院幼儿园设计方案时，当地反响如何？

马岩松（以下简称马）：2018年底，当我们发布概念设计时，不仅得到了幼儿园附近社区的关注，也得到了来自全北京甚至是海外的关注。很多家长都开玩笑说，如果采用我们的设计方案，他们愿意重上幼儿园！

赫：屋顶颜色的灵感来自于哪里？

马：在老北京，只有紫禁城的皇宫是彩色的，红墙金瓦，而大部分民居是灰色的。

我们希望新的建筑能与四合院保持适当、得体的距离，既不产生冲突，又能让人看到历史悠久的四合院。新的屋顶颜色应该与灰色的四合院区别开来，为什么要用红色？我想让它有一种来自火星的感觉。火星红让人有一种陌生但熟悉的距离感。

赫：设计时有没有尝试从儿童视角来想象事物？

马：设计这个幼儿园的时候，我一直在回忆自己的童年。我是土生土长的北京人，是在这种胡同环境中长大的。北京有句老话儿叫"三天不打，上房揭瓦"。四合院的屋顶就像院子的一道屏障，上了天台之后，你就可以克服这个障碍，翻到另一个院子去，或者到更高的地方去，离天空更近。孩子们都渴望自由，这就是童年经历给我的启示。

另外，四合院是一个古典空间，有着很强的空间秩序感。我们设计的建筑则构建了新的、自由的秩序。这种古典与自由的对抗与共存，说明我们需要生活在一个有秩序而不乏自由的空间里。对于孩子们的成长来说，这是很理想的环境。环境整体来说温馨如家，自由开放，这样会让孩子们更舒服。

赫：流动结构是否适合北京传统街区？

马：在胡同或四合院里，最重要的是规模，院子的大小、建筑的高低，都离不开人的体验。其次，四合院是一种内敛自足的空间，能够实现通风、自然采光和户外活动。只要把这两点维护好，我们就能实现新的建筑形式。

于是我们决定，幼儿园的新建筑一定要与老建筑不同，不过也要尊重老建筑在规模和空间类型方面的特点。因此，我们构建了一座单层建筑，为之设计了流动结构，并在下面搭配了几个庭院。这种新旧建筑之间的对话正是我们想要实现的目的。

赫：最后，我们能否通过建筑调和现代城市与自然的关系？

马：人工和自然并不是对立的，甚至自然也可以是人工的。现代城市中的大部分元素当然都是人工的。以古典园林为参照，我们能发现，即便是人工，也可以形成充满灵性的景观或场景。当然，现在的城市比起古老的园林要大得多，但我们仍然可以参考传统的园林理念，从中获得灵感。

What was the local reaction to YueCheng Courtyard Kindergarten when it was proposed?
In late 2018, when we released the concept design, it got attention not only from the community near the kindergarten, but also from people across Beijing and even from overseas. Many parents joked that they were willing to go back to kindergarten if it was this one!

What inspired the roof colors?
In old Beijing, only the Forbidden City's imperial palace is colored, with red walls and golden tiles, while most of the civilian residences were grey.
We hope the new architecture can maintain a proper and decent distance with the siheyuan. This should not generate any conflicts, but allow the historic siheyuan to be seen. The new rooftop color should differentiate it from the grey

siheyuan. Why red? I want it to feel as if it's from Mars. Martian red makes a strange but familiar feeling of distance.

Did the design try to imagine things through the eyes of children?
When I designed the kindergarten, I was always thinking back about my childhood. Because I'm a native Beijinger, I grew up in this kind of hutong environment. There is an old Beijing saying that "if you go three days without being punished, the roof will cave in". The siheyuan roof is like a barrier to the courtyard. After you go on the rooftop, you can overcome this barrier and go over to another courtyard, or go higher, closer to the sky. Kids long for freedom. That's what I got from my childhood experience.
Also, the siheyuan is a classical space that has a strong space order. The architecture surrounding it is a kind of new, free order. This confrontation and coexistence says we need to live in a space that is ordered, but also free. That's ideal for kids growing up. And the overall scale is cozy, just like home, free and open. This will make kids more comfortable.

Are fluid structures appropriate in traditional Beijing neighborhoods?

In a hutong or siheyuan, the most important thing is scale. The size of the courtyard, the height of the building – all relate to human experience. Secondly, the courtyard is inner/inward space that can realize ventilation, natural lighting and outdoor activities. As long as these two are well maintained, it can be the form of new architecture.
We decided that the Kindergarten's new architecture must be different from the old building, at the same time respecting it in terms of scale and space typology. Therefore, we built a single floor building, with several courtyards under fluid structures. This kind of conversation between new and old architecture is what we wanted to realize.

Ultimately, can we reconcile the modern city and nature through architecture?
The artifact and nature are not against each other. Even nature can be artificial. Nowadays in the modern city, of course most elements are artificial. If we take classical gardens as a reference, we do know that even artificialities can form some type of landscape or scene full of spiritual feeling. Of course, nowadays cities are much bigger than the old gardens, but we can still refer to traditional garden philosophy and get inspiration.

A-A' 剖面图 section A-A'

1. 教室 2. 庭院 3. 艺术教室 4. 行政管理办公室 5. 院长办公室 6. 办公室
1. classroom 2. courtyard 3. art classroom 4. administration office 5. principal's office 6. office

B-B' 剖面图 section B-B'

项目名称：YueCheng Courtyard Kindergarten / 地点：Beijing, China / 建筑师：Ma Yansong, Dang Qun, Yosuke Hayano / 设计团队：He Wei, Fu Changrui, Xiao Ying, Fu Xiaoyi, Chen Hungpin, Yin Jianfeng, Zhao Meng, Yang Xuebing, Kazushi Miyamoto, Dmitry Seregin, Zhang Long, Ben Yuqiang, Cao Xi, Ma Yue, Hiroki Fujino / 客户：Yuecheng Group / 执行建筑师：China Academy of Building Research / 室内设计：MAD Architects, Supercloud Studio / 景观设计：MAD Architects, ECOLAND Planning and Design Corporation / 功能：Kindergarten / 用地面积：9,275m² / 总楼面积：10,778m² / 高度：21.05m / 设计时间：2017 / 竣工时间：2020 / 摄影师：©Arch-Exist photography (courtesy of the architect) - p.8~9, p.11, p.12~13, p.15, p.16, p.17, p.19, p.20~21, p.22[upper], p.24; ©Hufton+Crow (courtesy of the architect) - p.22[lower], p.23, p.25

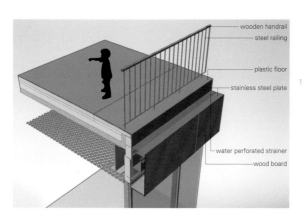

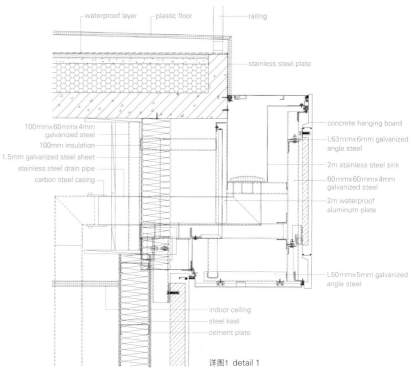

详图1 detail 1

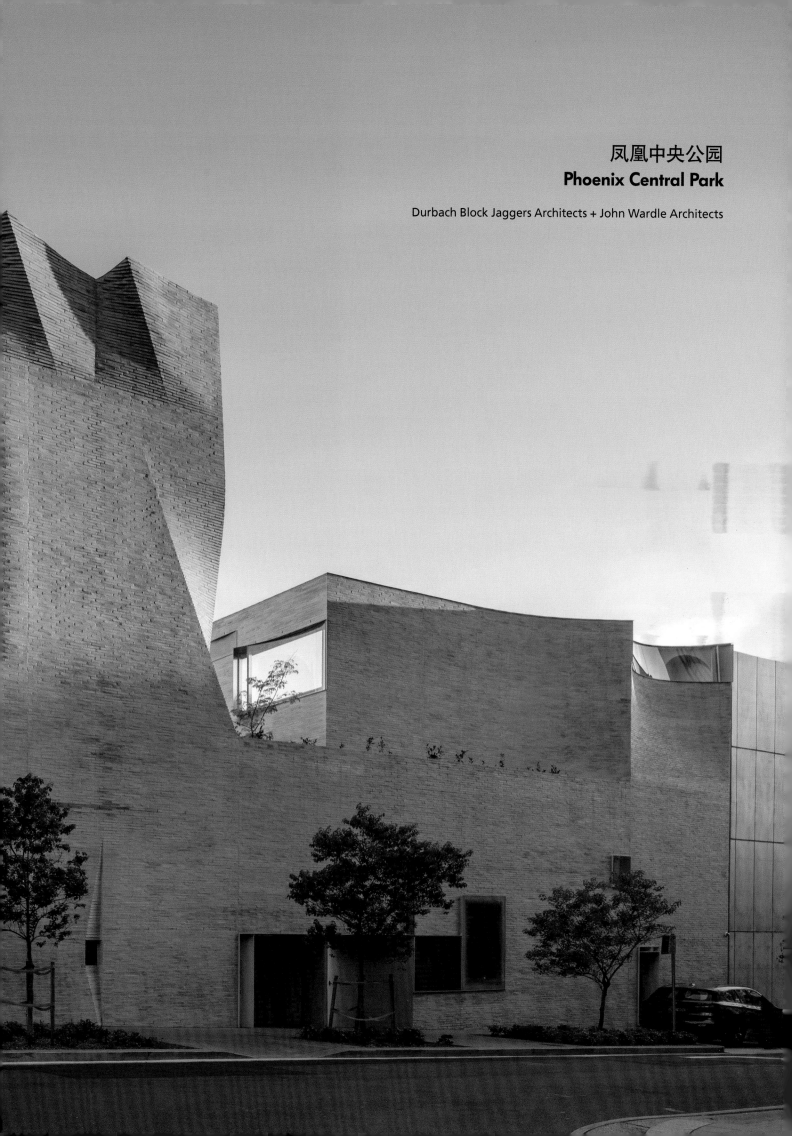

凤凰中央公园
Phoenix Central Park

Durbach Block Jaggers Architects + John Wardle Architects

建筑空间互联互通，砖墙立面富有雕塑韵律——艺术与表演融为一体
Art and performance are integrated in linked volumes wrapped in a sculptural brick facade

凤凰中央公园是一个全新的艺术中心，位于悉尼市中心的奇彭代尔社区，所处地块717m²，设有一处画廊和一个表演空间。画廊和表演空间分别由不同的建筑事务所设计，在材料和形式上采用了不同的建筑语言，但二者又彼此相连，共同组成一座1185m²的统一建筑。整个公园（包括庭院花园）实现了表演、自然和艺术的交织互联。除此之外，该建筑还采用了两家事务所合作设计的砖墙，有着雕塑般的质感。

画廊

画廊结构位于东面，由John Wardle建筑师事务所设计。建筑师对空间序列进行了编排，为人们提供了一场艺术之旅，从展示单个作品的私密独特房间，通往展示众多收藏品的广阔区域。画廊所在的东翼采用了混凝土墙，不同的空间复杂地堆叠在一起，并通过楼梯和桥梁彼此连接，保持了整体的鲜明特色。一楼大型画廊的地面由砖块铺设而成，砖面一直延伸到庭院。楼上的地板、楼梯和其他设施则为木材。

令人惊喜的景观、上方洒落的自然光和富有创意的楼梯更是引人入胜，让参观者移步向上。画廊屋脊棱角分明、鲜明突出，开凿了多扇天窗，它们给顶楼低矮却宽阔的空间带来的照明效果令人印象深刻，不仅避免了昏暗不明，而且还通过反射和过滤，将光线照射转化为柔和的光芒。

两扇面向北面街道的圆形窗户形成了收放自如的效果，它们位于外立面的平缓内凹处。这些窗户通过向室内凸出的墙体将画廊与外面的世界连接起来。庭院下方是一个小山洞般的空间，那里有一个朝上的天窗，让人们在静静地沉思作品的同时，也不会隔绝于上方白日里的热闹。

表演空间

Durbach Block Jaggers事务所设计的表演空间位于西翼，其高度低于东翼。它的中间是一个三层的钟形空间，由层层叠叠、流畅起伏的木肋构成。它的基底为地下一层，流线型的内表面受四面挤压向内集中，最终顶点形成了曲线长方形，四个角向外延伸。这一表演空间位于由多个门厅和通道构成的布局之中。像伊丽莎白剧院一样，表演空间位于布局的圆心部分，这意味着观众可以从舞台的周边看到表演。

建筑师在表演空间内设置了一圈略高于地面的凸出的楼座，从而形成了可替代使用的舞台或包厢。通道采用了直接布局或台阶，台阶的比例大小经过精心设计，观众可在此逗留，观看表演。建筑师还设置了一扇金色窗户，将街上的光照引入原本昏暗的空间。剧院空间采用的木衬板是用数字模板制作的。

剧院空间上方是一个有顶的艺术家会面空间，长边一侧通向一处空中花园，那里种满了多肉植物和树木。建筑外立面将顶层的这一设施与街道隔开，其中部分墙体高度较低，以容纳一扇弯曲的"云窗"，窗户的后面则是彩色拱形天花板。

外立面

画廊和表演空间由连续的砖墙外表皮连接在一起，砖墙也是整座建筑的外包层。两家建筑事务所经过了多轮对话讨论，反复磨合，合作设计了这一外立面。

砖块又长又平，采用叠石铺设方式，突出砂浆连接处。砖块上刷着一层薄薄的砂浆，凸显表面的连续效果，但在入口处周围，这种连续向内凹陷、扭曲，被切割并形成了拱形。

画廊面向街道的外墙向内挤压，形成圆形的凹陷，中心是大的圆环窗和较小的偏移窗。在它的上方，设置了天窗的屋脊倾斜的端部形成的屋顶轮廓线犹如海面上的波浪。临街一侧的砖立面形成了连接两翼的屏风，表演空间的砖立面为内退设计，立面上设计了长长的凹形曲线，最西端顶部则有一条"云窗"带。

画廊和表演空间的内部特质通过建筑外表面体现出来，同时，窗户、走廊和大门也为室内空间的设计增色不少。该项目是对一座建筑同时兼顾两种特性的一次有益的探索。

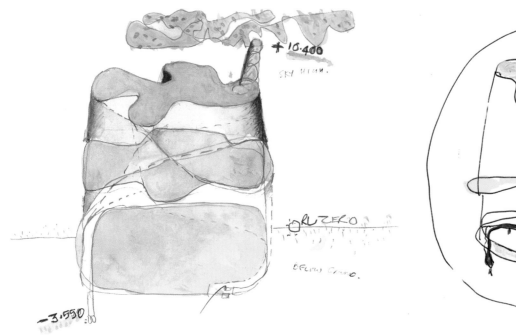

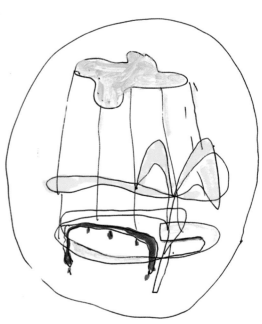

Phoenix Central Park, a new art center on a 717m² rectangular site in Sydney's inner-city neighborhood of Chippendale, contains a gallery and a performance space, each designed by different architecture practices. They have different languages of material and form, but are in dialogue with each other as linked wings of a unified 1185m² building. Including courtyard gardens, it enables an interlocking of performance, nature and art. The building has a sculptural skin of brick that was designed collaboratively.

The gallery

The gallery forms the east wing and was designed by John Wardle Architects. A sequence of spaces choreographs a journey from intimate, idiosyncratic rooms for the display of single works to expansive areas to showcase collections. With walls of concrete, this wing comprises a complex stack of differing spaces interconnected by stairs and bridges which maintain an awareness of the overall ensemble. A large ground floor gallery is paved with bricks which extend into the courtyard. Upstairs flooring, stairs and other features are of timber.

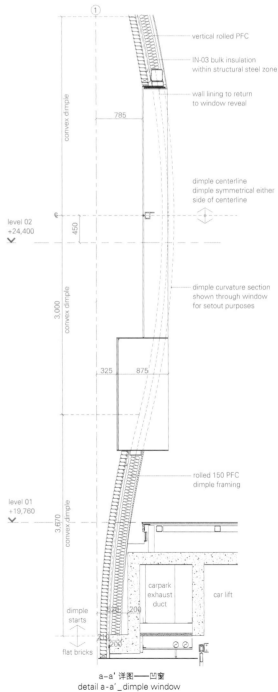

a-a' 详图——凹窗
detail a-a'_dimple window

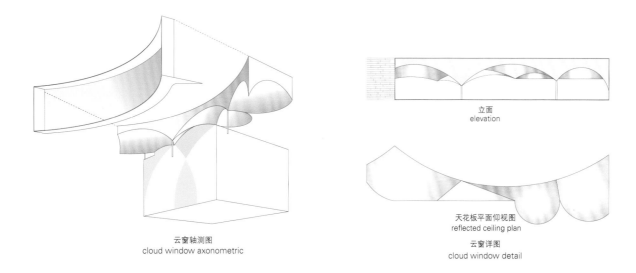

云窗轴测图
cloud window axonometric

立面
elevation

天花板平面仰视图
reflected ceiling plan

云窗详图
cloud window detail

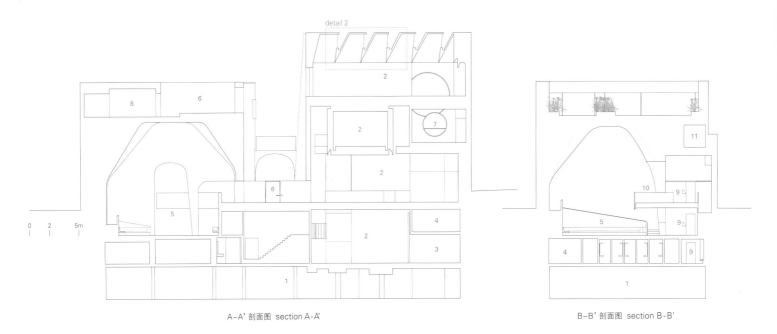

A-A' 剖面图 section A-A'　　　　B-B' 剖面图 section B-B'

项目名称：Phoenix Central Park / 地点：Chippendale, Sydney, Australia / 事务所：Durbach Block Jaggers Architects + John Wardle Architects / 画廊项目团队：John Wardle Architects–John Wardle, Stefan Mee, Diego Bekinschtein, Alex Peck, Luca Vezzosi, Adrian Bonaventura, David Ha, Ellen Chen, Andy Wong, Manuel Canestrini, Meron Tierney
表演空间项目团队：Durbach Block Jaggers Architects–Neil Durbach Camilla Block, David Jaggers, Simon Stead, Anne Kristin Risnes, Deb Hodge, Xiaoxiao Cai, Adam Hoh
项目经理：Aver 2015~2017; Colliers 2017~2019 / 规划方：Mersonn / 结构与土木工程：TTW / 岩土工程：Pells Sullivan Meylink / 建筑设备工程：Evolved Engineering
交通与人行道建模：GTA Consultants / 防火工程：Affinity Fire / 声学顾问：Marshall Day /

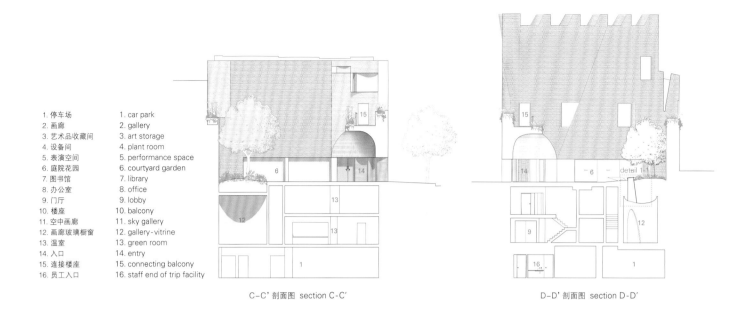

1. 停车场	1. car park
2. 画廊	2. gallery
3. 艺术品收藏间	3. art storage
4. 设备间	4. plant room
5. 表演空间	5. performance space
6. 庭院花园	6. courtyard garden
7. 图书馆	7. library
8. 办公室	8. office
9. 门厅	9. lobby
10. 楼座	10. balcony
11. 空中画廊	11. sky gallery
12. 画廊玻璃橱窗	12. gallery-vitrine
13. 温室	13. green room
14. 入口	14. entry
15. 连接楼座	15. connecting balcony
16. 员工入口	16. staff end of trip facility

C-C' 剖面图 section C-C'   D-D' 剖面图 section D-D'

景观建筑事务所：360° / 标识与指示牌：Studio Ongarato / 建筑测量与通道顾问：Philip Chun & Associates / 立面工程：Inhabit / 专业复合件制造商：Bluebottle / Specialty composite fabricators: Shapeshift / 承包商：Bellevarde Construction 2016~2018 (early works and superstructure); FDC Group 2018~2019 (structure and fitout) / 用地面积：717m² 总楼面面积：1,185m² / 建筑规模：6 levels - three above and three basement levels / 竣工时间：2019.12 / 摄影师：©Trevor Mein (courtesy of the architect) - p.26~27, p.29, p.32~33, p.34 lower, p.36 right, p.38; ©Martin Mischkulnig (courtesy of the architect) - p.30, p.34 upper, p.36 left, p.37 right, p.39 bottom, p.40; ©Tom Ferguson (courtesy of the architect) - p.31, p.37 left, p.39 top; ©Julia Charles (courtesy of the architect) - p.41

1. 画廊
2. 画廊上空空间
3. 门厅
4. 办公室
5. 浴室
6. 庭院花园

1. gallery
2. void over gallery
3. lobby
4. office
5. bathroom
6. courtyard garden

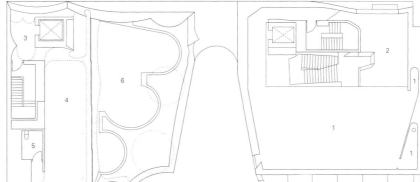

三层 second floor

1. 画廊
2. 图书馆
3. 楼座
4. 画廊上空空间
5. 连接楼座
6. 空中画廊
7. 表演空间上空空间

1. gallery
2. library
3. balcony
4. void over gallery
5. connecting balcony
6. sky gallery
7. void over performance space

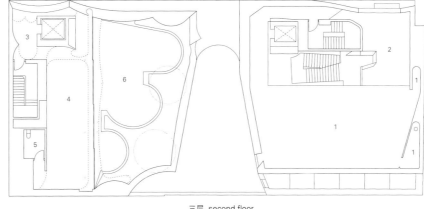

二层 first floor

1. 入口
2. 庭院花园
3. 门厅
4. 楼座
5. 表演空间上空空间
6. 画廊上空空间
7. 画廊
8. 升车机
9. 设备间

1. entry
2. courtyard garden
3. lobby
4. balcony
5. void over performance space
6. void over gallery
7. gallery
8. car lift
9. plant room

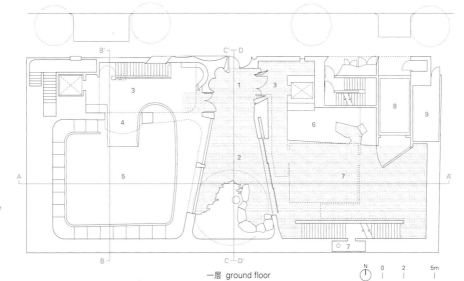

一层 ground floor

1. 升车机
2. 设备间
3. 画廊上空空间
4. 画廊
5. 门厅
6. 表演空间
7. 温室

1. car lift
2. plant room
3. void over gallery
4. gallery
5. lobby
6. performance space
7. green room

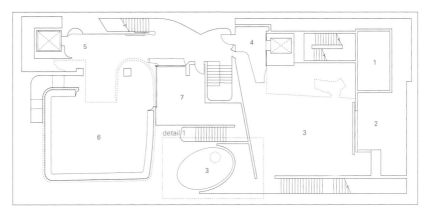

地下一层 first floor below ground

Unexpected views, natural light from above, and inventive stairs draw visitors upwards. A row of angular, sharp ridges carry skylights above the low wide space of the top floor, providing dramatic release from the darker spaces below. They reflect and filter the light into a soft glow.

A sense of containment and release is generated by two circular windows facing north to the street, in a gentle concave dimple in the facade. These oculi open the gallery to the world beyond, through a wall which is internally convex. Below the courtyard is a small cave-like space with an up-facing ocular skylight, offering quiet contemplation of a work whilst maintaining a link to the fluctuations of the day above.

Performance space
Durbach Block Jaggers designed the performance space in the west wing, lower in height than the gallery wing. It is centered on a three-story bell-shaped chamber made by stepped and contoured timber ribs. It rises from the first basement floor and its fluidic internal surface is squeezed inwards on four sides, so that its apex is a rectangle of curves with extended corners. This theatrical space is embedded in a fabric of lobbies and circulation spaces. Like an Elizabethan theater, the action is in the round, meaning the performance can be seen informally from around the stage.

A projecting balcony slightly above ground level loops into the space, creating an alternate stage or viewing box. The circulation is direct or via a set of stepped landings, scaled for lingering in and inviting overview. A gold window brings light from the street into an otherwise dark space. The theater is lined with timber fabricated from digital templates.

Situated above the theatrical void is a covered meeting space for artists, open on its long side to a sky-garden courtyard with succulent plantings and trees. This upper level facility is screened from the street by the facade wall, part of which is lower in height to accommodate a curving "cloud window" behind which are colorful arched ceiling structures.

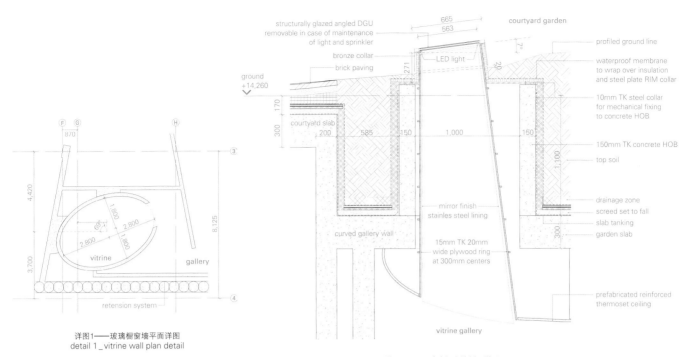

详图1——玻璃橱窗墙平面详图
detail 1 _ vitrine wall plan detail

详图1-1——玻璃橱窗墙剖面详图
detail 1-1 _ vitrine section detail

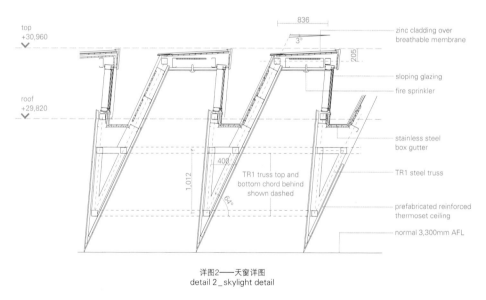

详图2——天窗详图
detail 2 _ skylight detail

The facade

The gallery and performance volumes are bound together by a continuous external skin of brickwork that encloses everything. The collaboration of the architects was iterative, and conducted with conversation and debate.

The bricks are long and flat, laid like stacked stone and emphasising the mortar joints. A thin veil of mortar over the bricks exaggerates the continuity of surface, which is dimpled, twisted, cut and vaulted around openings.

The gallery volume's street-facing surface has been pressed inward to create a circular dimple at the center of which is the large oculus window, and its smaller offset companion window. Above it, the inclined ends of ridges carrying the skylights form a roofline that is like waves on the sea. The street-side brick facade forms a screen bridging the volumes, and in the performance volume, is stepped back with lengths of concave curves, the western-most topped with a band of "cloud window".

Through this exterior surface the idiosyncrasies of each interior space emerge and erupt in windows, doorways and portals. It offers an exploration in how a building can be two things at once.

梅罗佩学校
Melopee School

XDGA

学校建筑与层叠的户外运动场地被统一在一个钢框架盒子结构内
Within a single steel frame box, a school building and stacked outdoor play areas are unified

在距比利时历史名城根特市中心约1km的地方,有一片狭长的码头。布鲁塞尔XDGA建筑事务所设计的梅罗佩学校和社区项目即位于此。该委托项目包括一所小学、一所幼儿园、一个课后辅助中心、一个食堂、一家咖啡馆,以及多处体育设施和操场。该建筑可容纳144名小学生和96名学前儿童,课后辅助中心可容纳56名儿童,另有一间日间托儿所可容纳29名儿童。该项目还具有社区性质,体育馆、食堂和会议室等设施可由学校和附近居民共享。

为应对空间不足问题、处理好项目内外复杂结构并开辟一条公共道路,建筑师将超大规模的建筑围护结构一分为二,然后再用20m高的镀锌钢矩形框架将其合而为一,形成整体。从水边到码头内道路,整个框架宽40m,高67m。其中一部分对应的是4630m²的紧凑实体建筑,它容纳了所有的室内功能空间,另一部分对应的是室外操场,面积为3050m²。梅罗佩这个名字来自于当地作家保罗·范·奥斯泰金所写的两位朋友同舟共济的故事。

框架内的建筑分为5层,高31m,外立面由不透明和半透明的聚碳酸酯板、玻璃和铝百叶拼接而成,非常牢固。幼儿园在一层,小学在二层和三层。孩子们登上楼梯就可来到教室,楼梯台阶黑白相间,犹如楼梯下摆放的那架钢琴的琴键。阶梯平台毗邻楼梯,师生可在此坐憩、站立和表演。而在建筑的西侧内立面所在的空间中向外望去,透过层层叠叠的操场,可以看到码头。四层是两层高的体育馆。五层设有一个带室外露台的咖啡馆,五层的东面有一个大看台,方便观众观看楼下的比赛。

室外结构从面向码头的外立面算起,进深为36m。钢结构的部分边框被钢网覆盖,随着植被在网内网外生长蔓延,建筑将被蒙上一层生机盎然、绿意盈盈的面纱。建筑师用玻璃砖构建了操场,下面是一条宽阔的公共道路,它穿过室外结构,沿着实体建筑设置。XDGA建筑事务所设计了各种游乐设施,如不锈钢滑梯、大型金属螺旋设施、攀岩墙、沙盒和一个半球形的坐井。从下向上看,这个坐井犹如倒置的穹顶。除此之外,还有一处倾斜花园,孩子们可以在其中照料菜地,菜地连接了整个建筑结构。室外结构南侧四层另有一处浮空篮球场。

整座建筑对声学性能的要求很高。建筑采用的矿棉板等吸声材料,上面覆盖着穿孔釉面砖。体育馆的面砖为蓝色,而食堂墙面的面砖为粉红色。虽然外部结构采用了严格的直线外形,但内部实体建筑则是一个精妙复杂、功能多样的多层结构,其户外延伸部分形状不同、配置各异,颇令人惊叹。梅罗佩学校的立体几何结构,无论是内部还是外部,都称得上是妙趣横生。

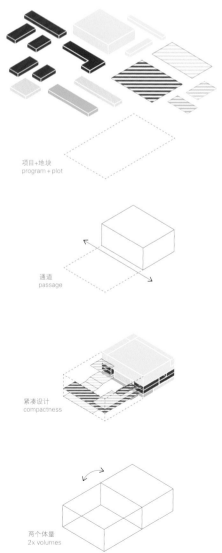

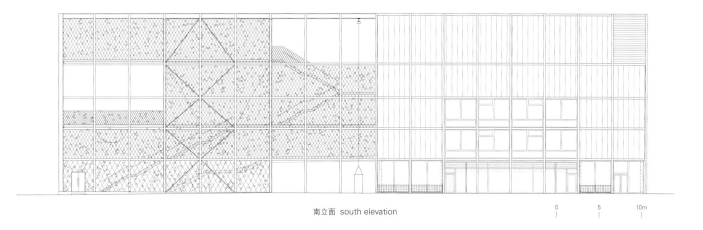

南立面 south elevation   0 5 10m

A narrow stretch of dockside, a little more than a kilometer from the center of the historic Belgian city of Ghent, is the site of a school and community project designed by Brussels based Xaveer De Geyter Architects(XDGA). The commissioned program for the Melopee School combines a primary school, a kindergarten, an after-school care center, a refectory, a café, sports facilities and playgrounds. The building can accommodate 144 primary school pupils, 96 pre-schoolers, 56 children in the care center and 29 in a day nursery. Because it is a community project, facilities such as the sports hall, refectory and a meeting room are shared between the school and its neighborhood.

In order to counter the lack of space, deal with the inside-outside complexity of programs and allow for a public path to pass, the maximum building envelope is divided in two parts. They are unified by a 20m-high rectangular skeleton of galvanized steel, 40m wide and 67m deep from the waterside to the inland road. One part is a compact solid building of 4,630m² housing all interior functions, and the other part is an outdoor stack of playgrounds within the skeleton frame, with an area of 3,050m². The name Melopee comes from a story about by Paul Van Ostaijen, a local writer, in which two friends share a canoe.

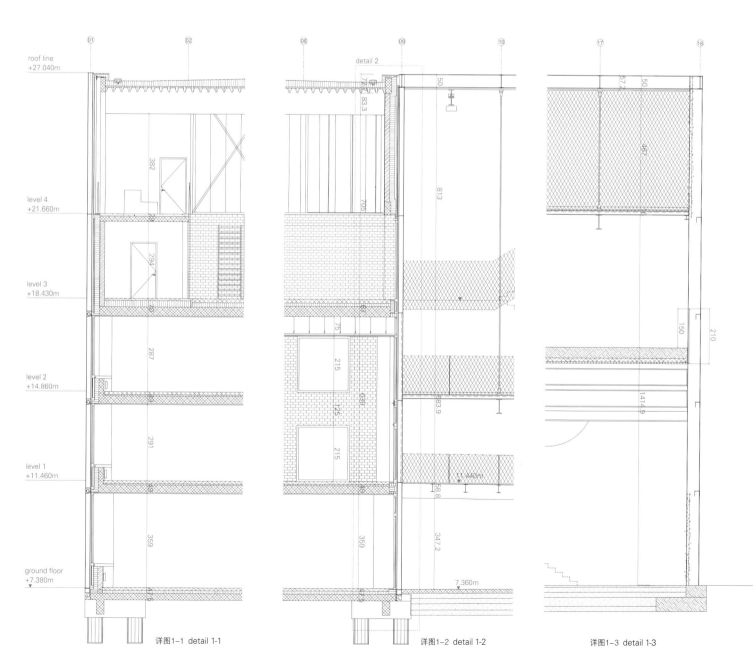

详图1-1 detail 1-1   详图1-2 detail 1-2   详图1-3 detail 1-3

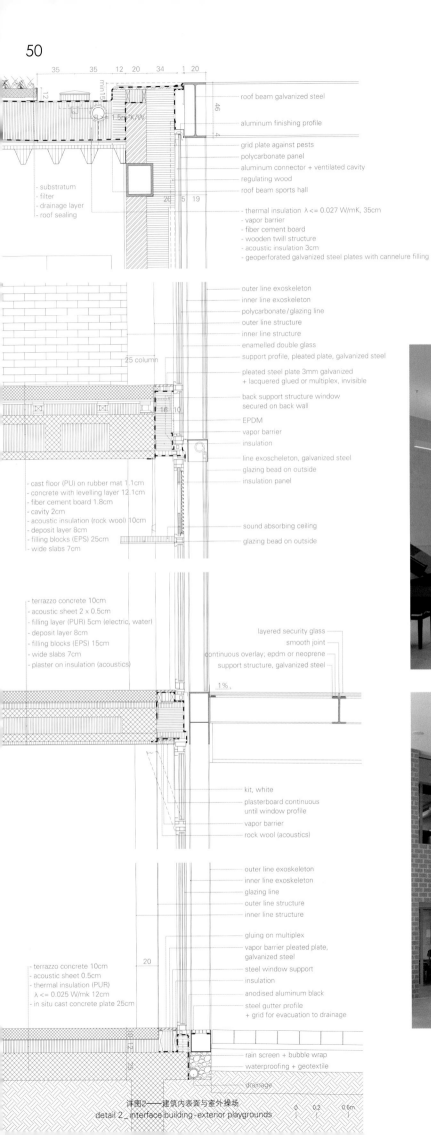

详图2——建筑内表面与室外操场
detail 2 _ interface building - exterior playgrounds

The 31m-deep five-story solid "interior" building has facades which are a patchwork of opaque and translucent polycarbonate, glass, and aluminum louvers. The kindergarten is on the ground floor, and the primary school is above, on the first and second floors. Children ascend to the classrooms upstairs which alternate between black and white, evoking a piano, like the one situated at the foot of the staircase. Adjacent to the stairs are stepped terraces for seating, standing and performance. Rooms on the interior western facade look out through the stacked playgrounds space towards the dock. A double-height sports hall on the third floor rises into the top level, which has a café with an outdoor terrace on the north side, and on the east side a grandstand so that spectators can watch the matches below.

The outdoor structure occupies a depth of 36m from the facade facing the dock. Parts of the perimeter sides of the steel structure are covered by a steel mesh, which will give the building a partially green, organic veil as vegetation grows across it. Under a playground realized in glass tiles, a wide public path crosses the outdoor volume beside the solid volume. XDGA designed the various play installations, which include stainless steel slides, a large metal spiral-shape, a climbing wall, a sandbox, and a semi-spherical sitting well which is visible as an inverted dome from below. A sloping garden, in which children can maintain vegetable patches, provides links across the structure. An outdoor basketball court on the south side is floated out at third-floor level.

The entire building generates high demands for acoustic performance. Absorbing materials include mineral wool panels covered by perforated glazed bricks, which are blue in the sports hall and pink in the refectory walls. While the external structure is strictly rectilinear, the solid building within it is an intricate layering of diverse functions, and the outdoor part reveals to the world a spectacular configuration of different shapes. The three-dimensional geometry of the Melopee School is playful, both inside and outside.

1. 操场
2. 运动场
3. 阶梯自助餐厅
4. 自助餐厅
5. 体育馆
6. 储藏间
7. 看台
8. 技术设备空间

1. playing field
2. sports field
3. terrace cafeteria
4. cafeteria
5. sports hall
6. storage
7. tribune
8. technical space

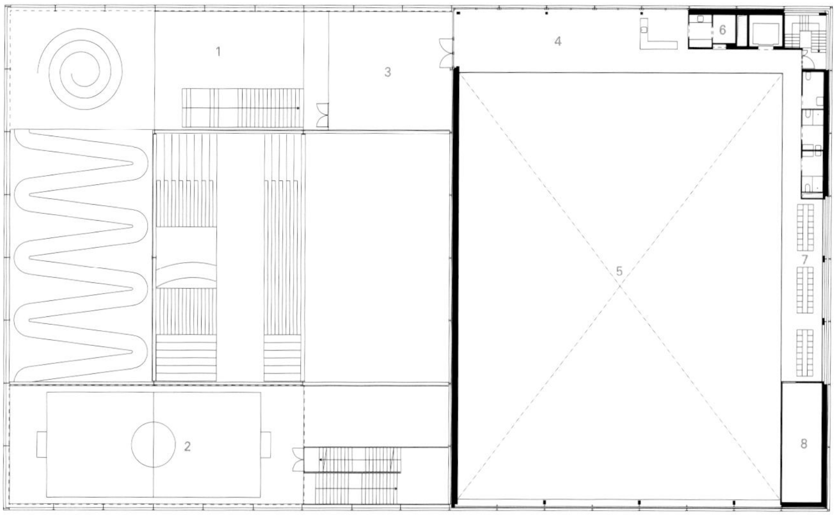

四层 third floor

1. 操场
2. 小学教室
3. 音乐工作室
4. 音乐理论教室
5. 储藏间
6. 食堂

1. playing field
2. primary school classroom
3. music studio
4. music theory classroom
5. storage
6. refectory

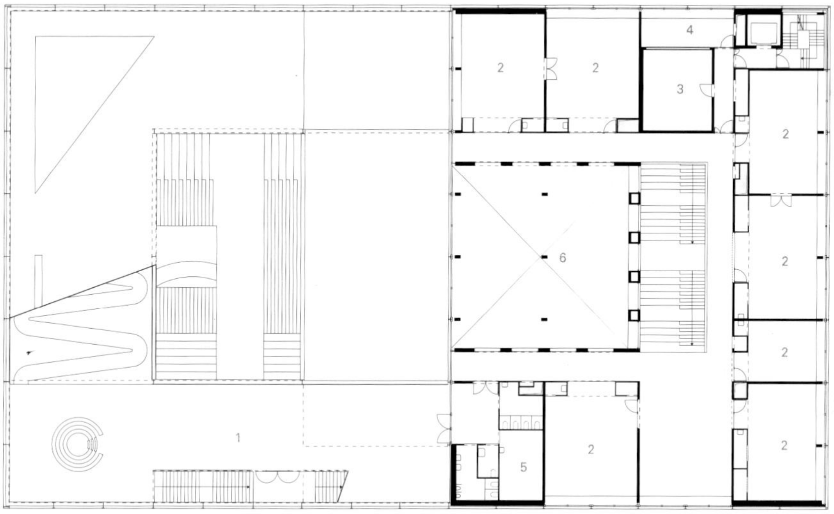

二层 first floor

1. 入口
2. 操场
3. 公共道路
4. 会议室
5. 员工室
6. 厨房
7. 办公室
8. 日托中心
9. 休闲空间
10. 室外空间
11. 秘书室
12. 技术设备空间
13. 储藏间

1. main entrance
2. playing field
3. public passage
4. meeting room
5. staff room
6. kitchen
7. office
8. daycare
9. relax space
10. exterior space
11. secretary room
12. technical space
13. storage

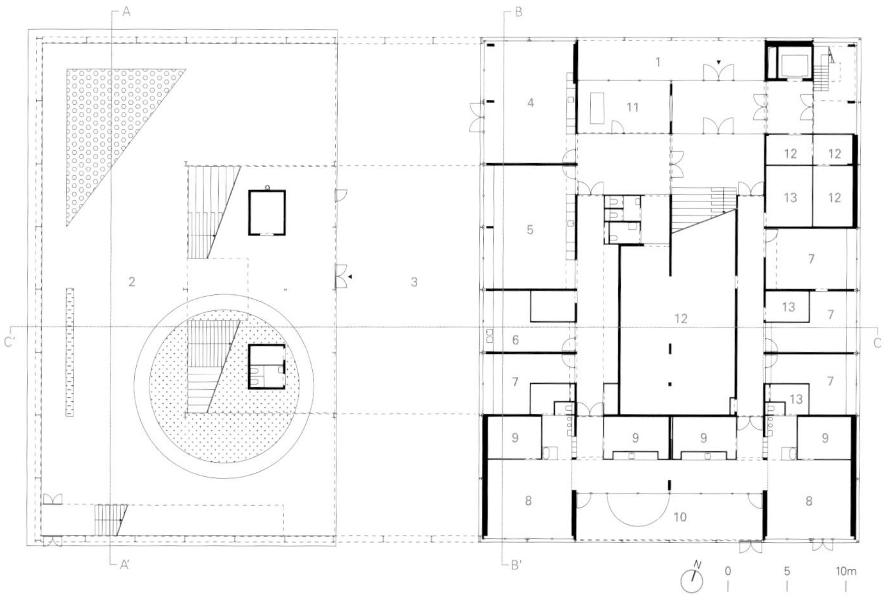

一层 ground floor

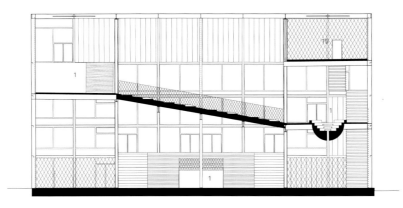

A-A' 剖面图 section A-A'

1. 操场	1. playing field
2. 公共道路	2. public passage
3. 厨房	3. kitchen
4. 技术设备空间	4. technical space
5. 办公室	5. office
6. 食堂	6. refectory
7. 小学教室	7. class primary school
8. 体育馆	8. sports hall
9. 看台	9. tribune
10. 会议室	10. meeting room
11. 员工室	11. staff room
12. 休闲空间	12. relax space
13. 日托中心	13. daycare
14. 自助餐厅	14. cafeteria
15. 运动场	15. sports field

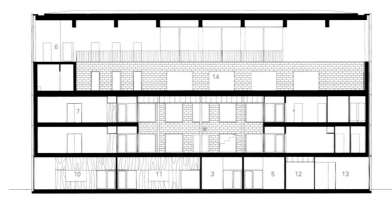

B-B' 剖面图 section B-B'

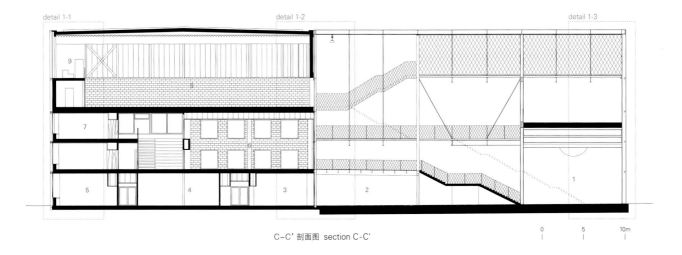

C-C' 剖面图 section C-C'

项目名称：Melopee School / 地点：Ghent, Belgium / 事务所：XDGA / 竞赛阶段项目团队：Xaveer De Geyter, Doug Allard, Thérèse Fritzell, Ingrid Huyghe, Willem Van Besien, Stéphanie Willocx / 最后设计项目团队：Xaveer De Geyter, Karel Bruyland, Thérèse Fritzell, Arie Gruijters, Ingrid Huyghe, Willem Van Besien, Stéphanie Willocx / 执行团队：Xaveer De Geyter, Ingrid Huyghe, Willem Van Besien / 顾问：Ney & Partners (structure), Studiebureau Boydens (mechanical), Daidalos Peutz (acoustics) / 客户：SO Gent / 功能：multipurpose school building / 用地面积：2,625m² / 总楼面面积：4,630m² (interior); 3,050m² (exterior) / 施工时间：2017.9—2020.2 / 摄影师：©Maxime Delvaux (courtesy of the architect)

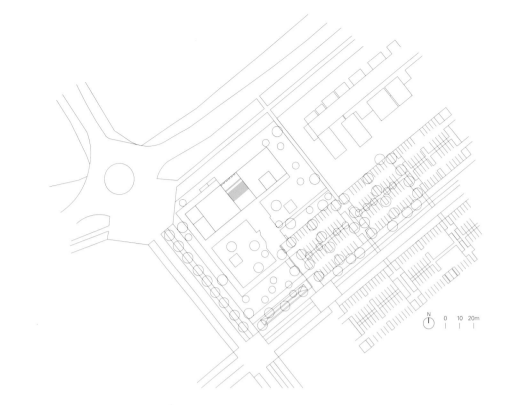

福特科技园
FORT Technology Park

RANDJA - Farid Azib Architects

混凝土墙围合的技术中心,设计灵感来自古城墙在战争中炸毁的小镇
A technology center within a concrete wall is inspired by an old war-damaged walled town

福特技术中心位于法国诺曼底圣洛市郊区一条环形公路旁,设有企业孵化器、联合办公空间、会议室/演讲厅和车间。该建筑综合体面积为1700m²,于2020年完工。顾名思义,福特技术中心如同堡垒般(注:英文名称Fort,意为"坞堡"),以坚实外墙围绕内部直线建筑和开放空间布局。建材选用和建造方法的灵感则来自第二次世界大战中被严重炸毁的圣洛古城墙。

技术中心的外周是长方形的围墙,内部则是两个底层相连的建筑主体。从外面看去,技术中心是一个长而有力的直线结构,结构表面采用了预制混凝土板,中间夹着一层保温层。西北侧的长外立面面向道路,从顶部垂直分劈为数段,直至底层。一层有一部分是玻璃墙体,上面是偏离建筑中心的深色盒子结构,该结构要高于外周墙体。

技术中心的东南面朝向乡村,一层设有多面全高的矩形石笼墙。这些石笼墙实际上是大门,可以围绕垂直中轴线旋转,它们是进入技术中心两个主体结构之间的中央空间的门户。门上钢丝网中填充着石块,与战时城镇遭轰炸后遍地的瓦砾相呼应。另外,还有两个方形天井通过这些大门向外开放。

中央空间最大的特点是纵向排列的台阶,这些台阶也可作为座位使用。台阶朝向西南逐级升高,最终到达两个主体结构之间的平台。在这里,人们可通过外围墙体之间的缝隙看到对面的道路。这一中央开放空间恰似围城中的公共空间。露台下面的连接空间包括一条通道、卫生间、餐饮和服务设施。封闭场地的西南端是一栋角落中设有天井的建筑。楼上是视频会议设施和会议室,亮度一般;高度更大的楼下则是一个拥有280个座位的礼堂,建筑师称之为"露天剧场",不过它既不像剧场那样采用曲线外形,也并非真的露天,而是两面都有玻璃幕墙。礼堂的座位呈阶梯状向西北方向下降。礼堂可以为附近的实习培训中心举办会议、研讨会和讲座提供场地,也可以为社区委员会提供会议服务。礼堂所在建筑比两侧的围墙高一层,但与围墙齐平。礼堂下方是一个大厅和大型演示区。室外台阶的线条向下延伸到一个玻璃结构的通道空间中,使得礼堂有一部分似乎漂浮在通道上面。

东北端延伸出来的双层箱形体量是商务中心。该体量在一侧的中心向内缩进,形成了另一处天井。商务中心高8m,是一个大型接待大厅。一层还设有一个休息室、一个协作式开放办公空间、一个由玻璃隔

南立面 south elevation

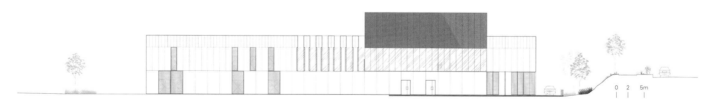

北立面 north elevation

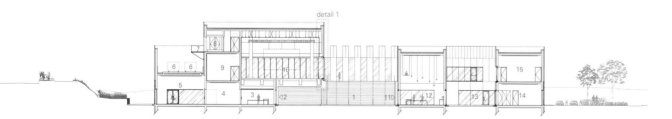

1. 入口前庭 2. 主入口——活动中心入口 3. 活动中心入口大厅 4. 缓冲空间 5. 展览室 6. 空气处理室 7. 消防员入口平台 8. 盥洗室 9. 视频会议室
10. 礼堂 11. 主入口——商务中心入口 12. 商务中心入口大堂 13. 天井 14. 车间 15. 创意室

1. entrance forecourt 2. main access-event center entrance 3. event center-entrance hall 4. buffer space 5. showroom 6. air handler 7. firefighter access platform
8. lavatory 9. video conference room 10. amphitheater 11. main access-business center entrance 12. business center-entrance hall 13. patio 14. workshop 15. creative room

A-A' 剖面图 section A-A'

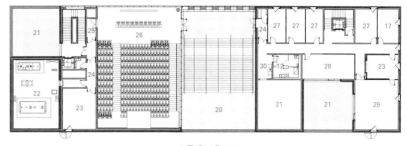

1. 前庭 — 1. forecourt
2. 天井 — 2. patio
3. 缓冲空间 — 3. buffer space
4. 门厅 — 4. lobby
5. 展览室 — 5. showroom
6. 储藏室 — 6. storage
7. 衣帽间 — 7. cloakroom
8. 接待处 — 8. reception
9. 锅炉房 — 9. boiler room
10. 垃圾处理室 — 10. waste disposal room
11. 餐饮区 — 11. catering area
12. 盥洗室 — 12. lavatory
13. 技术设备室 — 13. technical area
14. 联合办公经理办公室 — 14. co-working manager office
15. 档案室 — 15. archives
16. 储藏室 — 16. storage
17. 简单办公室 — 17. simple office
18. 开放布局办公室 — 18. open-plan office
19. 车间 — 19. workshop
20. 中央庭院 — 20. central forecourt
21. 上空空间 — 21. void
22. 空气处理站 — 22. air treatment station
23. 视频会议室 — 23. video conference room
24. 清洁间 — 24. cleaning room
25. 气闸仓 — 25. airlock
26. 礼堂 — 26. amphitheater
27. 双人办公室 — 27. double offices
28. 教室 — 28. classroom
29. 创意室 — 29. creative room
30. 休闲区 — 30. relaxation area

二层 first floor

一层 ground floor

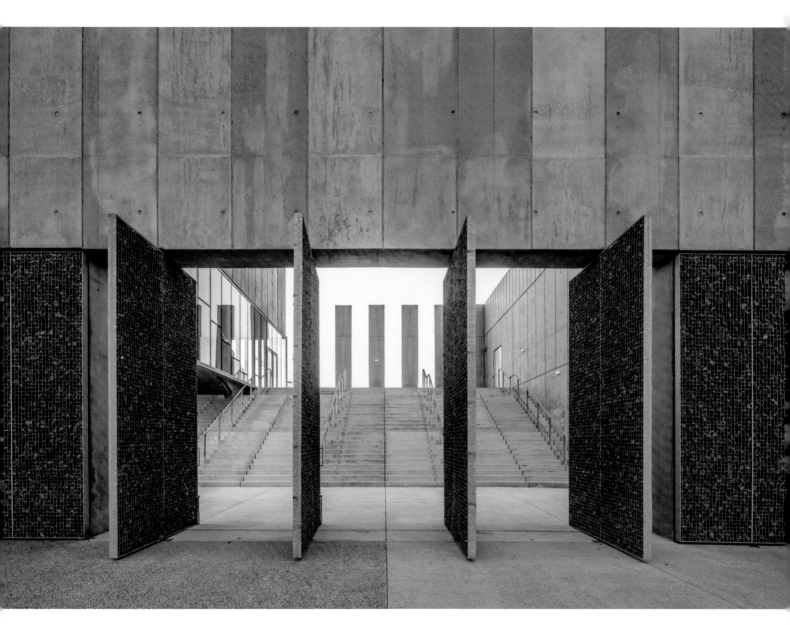

断的封闭式会议空间和一个FabLab车间。上层则是多间办公室、一间教室、一间大型创意室和一个小型休闲空间。

商务中心内部的地面、墙壁和天花板都由混凝土材料建成。天花板铺设了灰色吸声纤维板,标准高度为4m,使得室内空间更加宽敞。虽然外立面看起来基本上是不透明的,但从窗户进入室内的自然光十分充足,能够照射到建筑内的天井。天井也联系着不同空间之间的视角,人们可以从一侧看穿整座建筑。天井的灵感来自于中世纪城镇结构中的开放空间。

福特科技中心使用了2600m³的钢筋混凝土,建材来自20km外的一个采石场。混凝土在现场进行浇筑,板材也在现场组装。建筑内外都使用了混凝土,使建筑具有了强烈的野兽派风格。它的坚实风格也与法国传统建筑风格产生了共鸣——从简·普鲁维的作品到古堡和城镇的城墙,这种风格随处可见。虽然福特科技中心位于圣洛历史悠久的市中心之外,但它在许多方面都与圣洛遥相呼应。

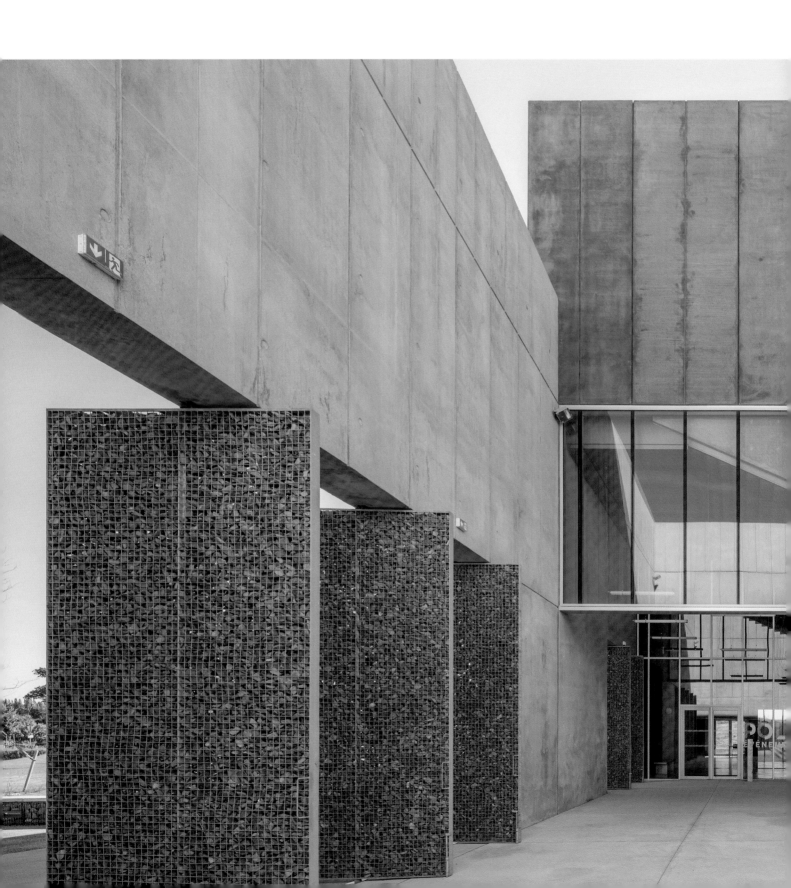

FORT is a technology center situated beside a road roundabout in the outskirts Saint-Lô in Normandy, France. Its functions include a business incubator, co-working spaces, a conference/lecture venue and a workshop. The name of the 1700m² building complex, completed in 2020, reflects its solid exterior wall around an inner composition of rectilinear volumes and open spaces, like a fortress. Its materiality and approaches to form were inspired by the old walled city of Saint-Lô, which was severely damaged by bombing in World War II. The perimeter of the building is a rectangle of wall containing two main volumes linked on the ground floor. The immediate impression from outside is of a long, solid rectilinear structure surfaced with precast concrete panels, which sandwich an insulation layer. The long north-west side facade facing a road is incised with vertical cuts from the roofline through the second of its two floors, and it is glazed below an off-center, darker box-structure that rises above the perimeter wall.

The south-east side faces countryside and on the ground floor includes a stretch surfaced with full-height rectangular gabion boxes. These are doors which rotate on a middle vertical axis

项目名称：FORT Technology Park / 地点：Saint-Lô, France / 建筑事务所：RANDJA_Farid Azib Architects
项目经理：Dhouha Hamdi / 助理：Isabelle Pinsolle, Yvanie Wilhem, Anouk Vialard
景观设计事务所：Christophe Gautrand et associés / 声学设计：LTE SAT
成本估算：VPEAS / 立面顾问：Elxir, Philippe Bompas
结构：Alpha Bet / 建筑设备：Artelia
客户：Saint-Lô Agglo
投资：Saint-Lô Agglo, Le Département de la Manche, La Région Normandie, L'Etat et La Communauté Européenne
经营者：Pôle Agglo 21 / 面积：1,700m² / 造价：EUR 4.2M
竞赛时间：2015 / 竣工时间：2020.1
摄影师：©Luc Boegly (courtesy of the architect)

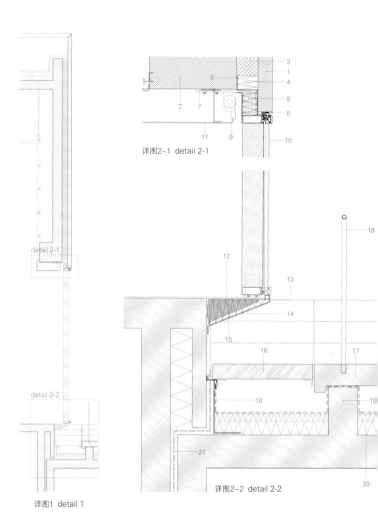

详图2-1 detail 2-1

详图2-2 detail 2-2

详图1 detail 1

1. rough concrete backing
2. architectural structural interior concrete wall
3. rigid thermal insulation incorporated into the casting
4. insulating stuffing
5. waterproofing membrane
6. water discharge flap including condensate water outlet
7. MR high spine fixing by sheet including wedging
8. spine edged in the upper part taken up by legs on a concrete structure
9. unguided manual roller blind including load bar
10. Clamaux curtain wall cladding of 3 sides seen by poly mirror stainless steel sheet
11. awning cover with removable elements in folded sheet metal, fixings by powder-coated screws
12. carpet covering "raw concrete" look
13. spine taken up in the lower part on folded sheet of great thickness as an extension of the interior concrete floor
14. interpon Futura (AkzoNobel) thermo-lacquered continuous cover including insulating padding
15. folded sheet with point reinforcements taken from the edge of the slab
16. concrete slab 7cm thick, supported by brackets
17. precast concrete staircase steps
18. waterproofing statement on curb
19. concrete rack beam
20. double layer thermal insulation
21. expansion joint

to become portals into the central space between the main volumes. The rocks filling their steel wire mesh echo with the rubble that filled the town after the wartime bombing. In addition, two square patios open to the outside with these doors. The central area is dominated by longitudinally-aligned steps which also serve as seating, and rise south-west to a terrace between the two main volumes. The terrace has views across the road through the gaps in the perimeter wall. This central open space is like a public space in a walled city. The link underneath the terrace contains a passage, toilets, catering and service facilities. At the south-west end of the enclosed site is a volume with a patio cut into its corner. Upstairs are video-conference facilities, meeting rooms, and below the darker, higher part of the building is a 280-seat auditorium which the architect calls the "amphitheater", although it does not curve, and is not open-air but curtain-wall glazed on two sides. It is designed with stepped seating descending to the north-west. It hosts conferences, seminars and lectures for the nearby Apprentice Training Center, or for community council meetings. This part of the volume has an additional story in height, rising above the perimeter wall on either side but flush to it. Under the amphitheater is a large hall and demonstration areas. The lines of exterior steps extend into a glazed circulation space, so that a part of the amphitheater seems to float over them. The two-story box volume extending from the north-east end is the business center. It is indented in the center of one side with the other patio. Under an 8m-high ceiling is a large reception hall. The ground floor also contains a lounge, an extensive open-plan office space for collaborative working, closed meeting spaces with glass partitions and a FabLab (workshop). The floor above hosts office rooms, a classroom, a large creative room and a small refuge for relaxation.

Inside, the floors, walls and ceilings are concrete. Grey sound-absorbent fiber panels line the ceilings, which have a standard height of 4m, contributing to the spaciousness of the interiors. Although the facades seem largely opaque, ample natural light enters the interior through windows opening onto the interior patios. They also articulate the perspectives between spaces, which can go through the whole building from one side to the other. The patios are inspired by open spaces in the fabric of medieval towns.

The building uses 2,600m^3 of reinforced concrete, using an aggregate from a quarry 20km away. Concrete pouring and panel assembly are carried on site. The use of concrete inside and out gives the building a strong sense of brutalism. Its robustness also resonates in French architectural tradition, from the work of Jean Prouvé to the ancient ramparts of fortresses and towns. While FORT lies beyond the Saint-Lô's historic urban center, it is in many ways an echo of it.

67

超级自然景观，为新世代而生

Hypernatural
for Future Ge

当人类干预自然时，只有形成环保的超级自然景观，才能实现人与自然的和谐共处。虽然超级自然景观属于人造景观，但在建筑的帮助下，它们对整体环境起到了巧妙积极的推动作用。因此，比起"人造"，叫它们"超级自然"更为妥帖。超级自然景观融会自然品质，并在此之上加入扩展或强化设计；而人造景观破坏自然品质、简化自然品质，更有甚者消灭自然品质。同是人类干预，超级自然景观保护环境，而人造景观则凌驾于环境之上。

Nature and people coexist seamlessly only when human intervention creates hypernatural landscapes that preserve the environment. Though made by humans, these landscapes have been helped along by architecture, with subtle nudges and energetic boosts. Calling them hypernatural, is more appropriate than calling them artificial. Whereas a hypernatural landscape takes the qualities of nature and adds to them, either by extension or intensification; an artificial landscape breaks the qualities of nature, renders them simplistic, or eliminates them at worst. Hypernatural landscapes preserve the environment after a human intervention, while artificial landscapes override the environment.

Lune de Sang展馆_Lune de Sang / CHROFI

山海美术馆_Mountain & Sea Art Museum / gad

周口店北京猿人洞1号考古遗址保护棚
The Protective Shelter of Locality 1 Archaeological Site of Zhoukoudian Peking Man Cave / THAD

艺术生物栖息地水景园_Art Biotop Water Garden / Junya.Ishigami+Associates

超级自然景观，为新世代而生_Hypernatural Landscapes for Future Generations / Phil Roberts

本文将列举四个人类干预产生的超级自然景观项目，其中有的是扩建，有的是强化，还有的是两者兼有。在这些项目中，建筑师都将自然环境放在与设计结构同等的位置上，甚至超过对建筑结构的考量。按设计方案，项目都尽可能地融于自然景观，甚至达到二者一体，而不是突兀显眼地脱离环境。此外，各项目以自己的方式恢复所在环境的生物历史，复原一度被遗忘的自然真实，更丰厚地回馈地球的赠予。

The four examples of hypernatural landscapes created by human intervention presented in this article can be seen as extensions or intensifications, and sometimes even both. In each project, the architects considered the natural context as much or more, than the structure that was designed. These are projects that have been designed to blend with the landscape as much as possible to the point where they are swallowed up by its presence, rather than making a rude statement by standing out. Furthermore, each project finds ways to reclaim the history of the site's biology, to bring into the present natural realities that were once forgotten, and to give back to the Earth more than what was taken.

超级自然景观，为新世代而生
Hypernatural Landscapes for Future Generations

Phil Roberts

建筑师仅将植被用于美化的时代已经过去。面对气候变化，人们需要新项目能自证其存在的合理性。超级自然景观能够恢复、保护项目所在地的生物多样性，扩大生物多样性益处，夯实巩固其适应未来的能力，造福子孙后代；公众也不再支持轻蔑自然景观的象征性建筑。于是越来越多的建筑师愿意在形式、结构、凝聚、和谐等方面遵循自然的引领，尽可能减少人类干预造成的影响。

以日本栃木县的艺术生物栖息地及其水景园为例(136页)，艺术栖息地于2019年建成，为艺术家提供了上好的休养地。那里的那须山山麓一片静谧，是艺术家寻找灵感的所在。在建造栖息地酒店时，石上纯也事务所本想大量砍伐林木，但虑及可能引发的争议，于是决定花四年时间，将整个森林连根拔起，移植到毗邻酒店的沼泽地上。另外，事务所将草地改造成超级自然水景园，就此改写了当地的稻田用地史。园内池塘由当地已有闸门放水而成，散布在移植后的树木间，均匀反射阳光，让阳光照满整个水景园。池塘必须蓄水不溢，因为被移植的树种(栎类、犬蔷薇和山毛榉)不能临水种植。318棵树和160处池塘之间长满青苔，垫脚石铺成的步道蜿蜒伸展，引导客人走向水景园。如今，贯穿当地历史的植被和水体交相层叠，融为一体，恢复昔日环境，为我们带来强化形式的超级自然景观。

建筑师对此解释道："如果按照自然规律，这种水景园是无法实现的，因为水池会抬高水位，造成树木无法生长。事实上，这座水景园是一件精心设计的艺术品，它掩盖了使之形成的人工元素。"艺术生物栖息地水景园具备自然景观之美，并在此基础上将其发扬光大。这里为访客带来了无与伦比的体验，让他们感受情感上的联系，徜徉其中

Gone are the days when architects would add vegetation for aesthetic purpose only. Climate change has meant that the population demands that new projects justify their existence. With hypernatural landscapes, we restore or preserve a site's biodiversity, intensify its benefits, and make them resilient for future generations. The public will no longer stand for symbolic architectural gestures that disrespect the natural landscape. More architects are willing to follow the lead of nature in terms of form, structure, cohesion, and harmony in order to minimize the impact of a human intervention.
Consider the Art Biotop in Tochigi Prefecture, Japan and its Water Garden (p.136). Completed in 2019, the Art Biotop is a retreat for artists, where they can seek inspiration in the serenity of the foothills of the Nasu Mountains. To build the hotel, Junya.Ishigami+Associates had to cut down numerous trees in a forest, a move that would have been controversial. Instead, the architects decided to uproot the entire forest and transplant it onto the marshy site adjacent to the hotel site over the course of four years. Transforming the meadow into a hypernatural water garden, the site's history as a paddy field was recreated. Ponds filled with water from an existing sluice gate are interspersed between the transplanted trees to reflect sunlight evenly across the garden. The ponds have to be waterproofed because the transplanted species of trees (Quercus, canine, and beech) cannot live near water. Between the 318 trees and 160 ponds, moss is applied and circuitous paths of stepping stones give guests access to the water garden. The vegetation and the bodies of water that were on the site throughout history are layered on top of each other in the present day to create this hypernatural landscape in the form of an intensification by restoring the past.
As the architects explained, "such a garden should be physically impossible because the water basins would raise the water level to the point that the trees couldn't thrive. This water garden is, in fact, a carefully calculated work of art that masks the human element that made it possible." The Art Biotop Water Garden takes the beauty of the natural landscape and enhances it. It is a place where guests are drawn to

犹如梦中漫步，阳光照射在众多池塘上，池塘反射的微光覆盖着大地。这座堪称奇幻王国的水景园不仅为我们带来了自然界难以实现的强化设计效果，也为我们带来了近乎人造环境下非科技沉浸式体验的享受。

自然景观固然设置了生物模式，但超级自然景观的能力却要更胜自然景观一筹。在北京的北京猿人洞，一个珍贵的考古遗址正面临着被毁的危险。二十多年来，气候侵蚀一直在威胁着这个联合国教科文组织的世界遗产中心。清华大学建筑研究院建议扩建一个保护棚，这样既可以将其他人造物挡在外面，又可以让阳光和空气进入洞内（118页）。洞内树木仍然能够享受阳光，还能接受精心安排的降水。有了这个保护棚，就可以调节洞穴内部温度，让游客参观游览，在洞内举办活动，而且全年都可以使用。从远处看，这个保护棚看起来像是山区景观的一部分，但从近处看，洞穴覆盖面是一个由825块独立面板（建筑师称之为"叶片"）组成的屋顶。叶片均有一个向外朝向天空的绿顶和一个向内的岩石面。从里面看，保护棚似乎是洞穴与生俱来的一部分。

创建这个超级自然景观项目，离不开对气象数据的应用。清华大学建筑研究院分析数字风洞技术，模拟保护棚高度、叶片间隙、温度、日光、通风和其他变量，通过计算收集数据。结果，保护棚呈现有机形态，可随景观滚动弯曲。多年后，随着叶片屋顶上的植被生长，整个保护棚将在周边的树冠中隐介藏形。同时，叶片彼此之间仅相隔几厘米，但足够让洞内进行充分的空气流通。阳光会在一天中的不同时间，偶尔穿透到保护棚下的岩缝中。

当建筑师面对地形束手无策，只能基于当前凹凸不平的斜坡进行设计时，他们往往会为这种自然景观采取超级

feel an emotional connection through an experience that is incomparable. Walking through the garden is like meandering in a dream, with the ground shimmering with sunlight from the many reflective ponds. The garden is a place of fantasy, because rarely in nature does such intensification exists. It is the closest thing to a nontechnological immersive experience in a manmade setting.

Whereas natural landscapes have set biological patterns, hypernatural landscapes can be built to do things their native counterparts cannot. At the Peking Man Cave in Beijing, a precious archeological site was at risk of being ruined. For over two decades, weathering was threatening this World Heritage Center of UNESCO site. Architecture office THAD proposed preserving the cave by building an extension in the form of a protective shelter that would keep the elements out, while still allowing sunlight and air in (p.118). Within the cave are trees that still benefit from the sun, and are watered in a controlled manner. The protective shelter allows the temperature to be regulated inside the cave, allowing for visitors, activities, and programming 12 months of the year. From afar, the protective shelter looks like part of the mountainous landscape, but up close, the covering is a roof made of 825 individual panels, or as the architects call them, blades. Each blade has a green roof facing the sky, and a rock face on the interior. On the inside, the shelter looks as if it has always been there.

To create this hypernatural project, meteorological data were used to give form to the shelter. THAD describes the process of gathering data from computations by analyzing numerical wind tunnel techniques, to simulate the height of the shelter, gaps between blades, temperature, humidity, daylight, ventilation and other variates. As a result, the protective shelter takes on an organic form that rolls and curves with the landscape. Over the years as the vegetation on the roof of the blades grows, the whole shelter will disappear into the surrounding tree canopy. At the same time, the blades are separated from each other by only a few centimeters, which is enough for adequate circulation in the cave. Depending on

自然景观项目。中国昆明的山海美术馆(100页)的设计就模仿了陡峭的山坡。在最初的构想阶段，总部位于中国的gad建筑公司并不想建一座单层美术馆，因为这需要为基础开辟更大的区域，会破坏整体景致。所以他们决定在山顶建造一座断环式莫比乌斯带建筑，让石林插入环带建筑断裂处，成为美术馆内部布局的一部分。

正如建筑师所说，地形是"激发设计灵感的先决条件"。基于此，美术馆成为山丘的自然延伸，建筑设计更是凸显了山坡的崎岖不平。除此之外，美术馆的建筑结构还有很多特殊之处：墙壁锐角夹角、天花板下沉、窗户倾斜，人们在馆内即可看到室外的树木和岩石。除此之外，建筑师还巧妙利用了红黏土和夯土的特性，使得室内更像是一个宽敞明亮的洞穴，斑驳的阴影落在浅色的墙壁上，大理石地板在强烈的阳光下熠熠生辉。美术馆依山而建，顺势延伸，这种对山体的夸张复制创造出一种超级自然景观。整座建筑层层叠叠、动感十足，各部分看似相互冲突，彼此却又紧密相连，相互呼应，与周围的自然环境唇齿相依。但是，这样的设计也存在问题，那就是如同这个斜坡本身一样，依坡而建的美术馆也存在不稳定性。根据建筑选址规则，美术馆内部采用的冷静风格应在物理和外形上保持一定的稳定平衡，避免滑入怪异设计的误区。自然和人为因素融为一体才可以成就超级自然景观。

澳大利亚新南威尔士州北河区的Lune de Sang展馆项目(76页)则是人工干预的成功范例。展馆高度很低，接近地面，让人很难注意到。从某些角度看，它的石墙似乎与景观融为一体。展馆占地141.6ha，用地几经更迭，还曾做过乳牛场。CHROFI建筑师事务所用当地特有的材料(部分甚至是回收材料)设计了这座低碳展馆。这座展馆在岁月中逐

the time of day, sunlight manages to sneak its way sporadically down onto the crevices of rock below. When the topography does not give the architects any choice but to design with an existing jagged slope, the project gets grafted into the landscape. In Kunming City, China, the Mountain & Sea Art Museum (p.100) mimics a steep mountainside. The architecture firm, China-based gad, did not want a museum all on one level, because that would have required disturbing the landscape by carving an area for a foundation. Instead, they decided to lay the building out like a fractured Möbius strip on top of the mountain, permitting slabs of stone to pierce through to the interior.

As the architects described, the topography became "a prerequisite for triggering the design." The museum becomes an outgrowth of the mountain, with the architecture of the building reflecting the unpredictability of the slope. Walls clash at acute angles, ceilings dip, windows slant, while trees and stone are seen within the building. With building materials such as red clay and rammed earth, the feeling inside is that of a well-lit cave. Irregular shadows touch lightly colored walls and marble floors glisten under intense sunlight. The architecture creates an extension by taking the precipitous forms of the mountain, and copies them in an exaggerated way to create a hypernatural landscape. This makes for a dynamic site on multiple levels, all seemingly in conflict with each other, but undeniably connected. Every portion communicates with the next, while they all communicate with the surrounding mountain and trees. The design is erratic, just like the surfaces of the mountain, and that is the point. This is a museum where the calmness of the interior spaces is physically and figuratively balancing on the edge of eccentricity because the architecture has chosen to follow the rules of the site. Having the natural and the manmade feel like one is what makes this a hypernatural landscape.

Lune de Sang (p.76) in the Northern Rivers region of New South Wales, Australia, is an example of a human intervention that is both an intensification and an extension. The pavilion is so low to grade that it is barely noticeable, and from certain perspectives its stone walls appear to be part of the landscape. This 141.6ha site was once a dairy farm, but that is only one aspect of its history. Built with materials vernacular to the

渐成熟，最终犹如精心保存的废墟一般。它成为森林中央的配置项目，而这片森林最终将成为可持续开采的森林。展馆所在场地至少种植了10万棵阔叶树；在未来的50到300年中，这些树木将逐渐成熟，形成茂密的森林，届时展馆就可以隐匿其中，与自然融为一体，完全不会被人注意到。

　　建筑师John Choi在公司的YouTube短片中提到，"整个景观经历了悠然绵长的时光，Lune de Sang展馆项目的灵感即来源于此。"历经120年的农耕砍伐之后，该地区土地退化严重，建造这个超级自然景观对修复该地区的自然环境有莫大的助益。项目的长远性使得建筑师不得不考虑数百年跨度的项目生命周期。未来，森林将会成为超级自然景观的主体，而森林之下的建筑都将隐匿其中，与之融为一体。

　　就上述四个项目而言，自然环境是建筑发展的一个决定性因素。因此我们完全可以大胆向大自然取经，深入探究形式、结构、凝聚力以及和谐这四个方面对于超级自然景观的重要意义。在这些项目中，建筑师将大自然教授的东西牢记于心，让设计服从于自然。这就要求建筑师将景观放在第一位，而不是一味地追求建筑本身。在超级自然景观中，建筑能很好地融入自然，而不是如以前一般与自然抗衡，喧宾夺主。每个项目都展示了其尊重自然环境的内核。在这些项目中，不仅现有景观得到了尊重，它们的自然历史也得以再现，有时甚至可以将过去的一切一幕幕地还原。这些自然环境之所以得到了成功的扩展强化，原因就在于对超级自然景观的运用。这说明，回报地球的最好方式就是将建筑微妙地融入自然，甚至用建筑来为自然增色。

area, some even recycled, CHROFI architects designed a low-carbon pavilion that would itself mature over time like a finely kept ruin. A fitting project in the middle of what will become a sustainably-harvested forest. At least 100,000 hardwood trees were planted on the site, and over the next 50 to 300 years this intensification will mature into a rich forest making the pavilion completely unnoticeable.
"The slow timeframe of the landscape," said architect John Choi during a short film on the firm's YouTube channel, "has really been the inspiration for the project." The creation of this hypernatural landscape was imperative to heal the site's degradation after 120 years of farming and logging. This project corrects what was taken away from the site and restores the natural context that was there prior to the land being industrialized. Taking the long view of the project forced the architects to think in terms of a life cycle that would span hundreds of years. In the future, the forest will be the dominant feature on the landscape, while the pavilion and its sheds will all be tucked away.
For all four of these projects, the natural context was a determining factor in the development of the architecture. Nature has much to teach us about form, structure, cohesion, and harmony, but we need to be willing to learn the lessons. In these projects, the architects took those lessons to heart, and made design submissive to nature. This could not have been achieved if the architecture had been the number one priority and the landscape treated as secondary. With hypernatural landscapes, the human intervention is the secondary element, and to do that requires architecture that is humble, not architecture that boasts and makes a claim on a territory, making a rude statement with its presence. Each project presents architecture that is respectful to its natural context. Not only are the existing landscapes respected, but their natural histories are reanimated, sometimes bringing layers of the past into the present. These extensions and intensifications of the natural context are only made possible because the architects seek to create hypernatural landscapes that blend with the human interventions. In doing so, they have proved that the best way to give back to the Earth is not to disturb it, but to add to it with subtle architectural gestures and sometimes even energetic boosts to its biodiversity.

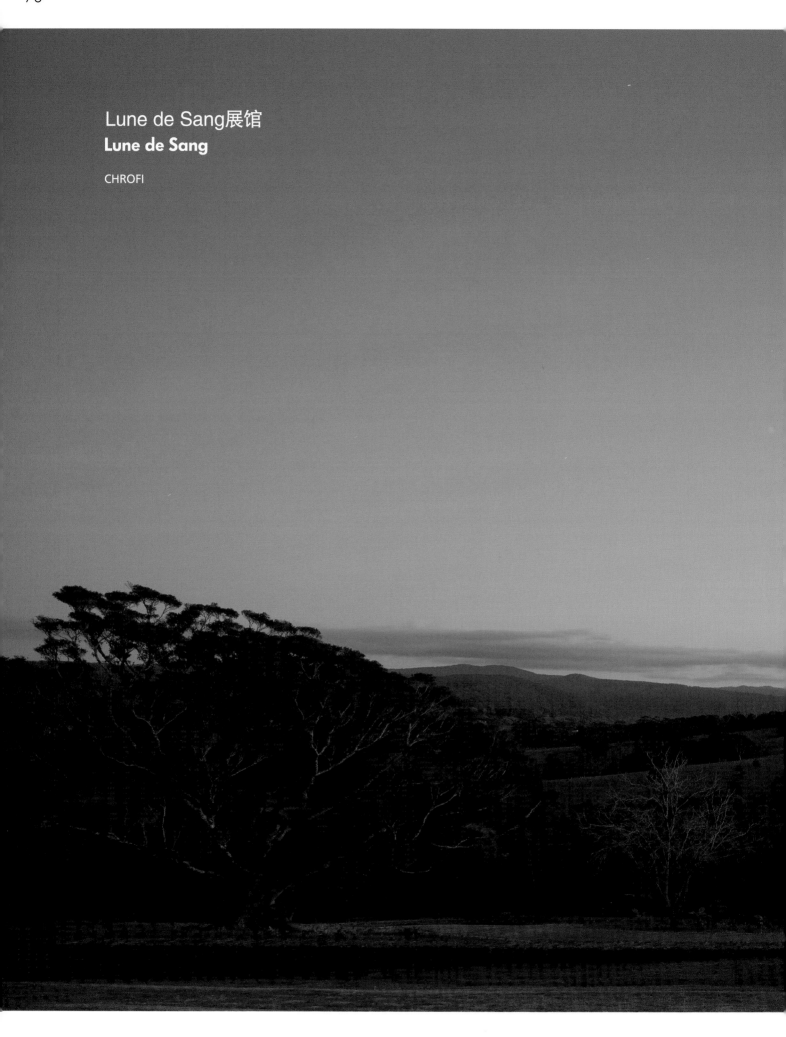

Lune de Sang展馆
Lune de Sang

CHROFI

基于阔叶树生长时间的人工林中的三个结构设计
The growth time of hardwood trees informs the design of three structures in a forest plantation

在距离澳大利亚新南威尔士州拜伦湾25km的内陆地区，92ha的土地因奶牛养殖而退化。不过，现在这里已经重新种植了当地的热带雨林阔叶树种。尽管其中某些树木需要300年才能成熟，但是它们终将会给这片土地带来收益。Lune de Sang展馆项目较长的工期说明了一座展馆和两个工作棚建筑设计的特别。悉尼建筑事务所CHROFI以废墟为切入点，展望长久、进化和自然之间的交织，为建筑选择了混凝土材料。它会随着时间的推移而成熟，进而强化建筑质量。

展馆占地280m²，分为两个部分：一个是开放的、可以远眺的公共空间，另一个是隐蔽的私人居所。展馆公建位于混凝土梁框架支撑的矩形屋顶下。长边侧的水平横梁悬在挡土墙顶部旁，并承载着一排天窗框架，将屋顶向上抬升，使其倾斜，对面则设置了三面落地玻璃，封闭了两根垂直混凝土支柱之间的空间。这一设计就如同将森林引入了室内空间。长长的横梁和屋顶从封闭空间中延伸出来向外悬挑，外部水平横梁的一端延伸到露天游泳池上，游泳池的外壳则嵌在石墙内，此处的土地是下陷低于地面的。

从物理上说，这座建筑是一个单一结构，但要从公建进入私人居所，就必须通过挡土墙上将两个区域分隔开的门。私人居所内镶嵌在混凝土框架内的全高玻璃墙使得这里成为一栋玻璃房。在这里不能一览无余地看到周围的全景，这里面对的是苍翠繁茂的雨林堤岸景观，因此空间拥有一种地下空间的品质。私人居所内的空间隔断采用了细木业常用的黑基木（一种桉树），而屋顶则由从挡土墙上以不同角度延伸而出的两个方形区域结合而成。在其中较长的一个方形下，卧室和浴室彼此相连。私人居所的比例和规模十分适宜，显示出精湛的工艺水平，确保该建筑的使用寿命可达数个世纪之久。

展馆是一栋可持续建筑。混凝土与石材结构是蓄热体，可以保持空间凉爽；使用后形成的中水最终返回森林生态系统。该展馆设计在确保碳成本最小化方面也投入了大量精力，使用本地资源和可回收材料就可以说明这一点。例如，石墙就是用从周边森林中开采的岩石建造的。

Lune de Sang两个工作棚建于展馆之前。它们与土地融为一体，设计富有韵律，土坡和石质挡土墙上鲜明的元素重复设计尤为律动。工作棚都建在山坡上，在混凝土与泥土蓄热体的加持下保持了冬暖夏凉。

1号棚占地320m²，由从石墙路堤延伸出的平行混凝土单元构成，并且安装有条形照明灯。平行混凝土单元垂直折转向下，形成19根在其外侧的平板支柱。安静、隐蔽的棚屋空间由三面玻璃墙封闭，玻璃结构在两侧分别让出三根平板支柱的位置。

在工作棚主空间的后方，还有一个伸入山坡的工作间。游客通过工作棚的结构框架，穿过富有韵律的支柱向前望，能够看到葱茏的树林，举目远眺，还可以看到外面开阔的围场和无花果树，以及蜿蜒伸入建筑的石墙。

2号工作棚占地360m²，与1号棚理念相反，采用了开放式结构。向上倾斜的镀锌屋顶为下面的储存空间提供了遮蔽，而11根20m长的平行混凝土屋梁则从山坡穿出，固定了11m长的部分悬挑的屋顶。在屋顶下，围合矩形空间的木墙在正面可以向上掀开，在侧面可以像门一样打开，大大方便了材料的搬运和储存。木墙关闭时，由于没有和屋顶相连，2号工作棚空间仍然是开放的。

无论在何时，人们只有抵达两个工作棚近处才会发现它们，这种设计反而强化了它们在景观中的存在感。

25km inland from Byron Bay in New South Wales, Australia, 92 hectares of land degraded by dairy farming have been replanted with indigenous tropical rainforest hardwood species. The trees will be harvested, but some will take up to 300 years to reach maturity. The deep timescale of the Lune de Sang project has informed the architecture of a pavilion and two working sheds. Sydney-based practice CHROFI looked at ruins to envision the intertwining of permanence, evolution and nature. Concrete was chosen because it would mature over time, intensifying the buildings' qualities.

The 280m² Pavilion is divided into two parts: an open, communal gathering space with long views, and a hidden private retreat. The communal component of the pavilion lies under a rectangular roof supported by a concrete beam frame. The horizontal beam of one long side floats beside the top of a retaining wall, and carries a line of skylight window frames that push the roof up, so it tilts. On the opposite side, three sides of glazing enclose a space between two vertical supporting concrete columns. This floor to ceiling glass brings the forest inside the space. The long beams and roof cantilever out from the enclosed space, and on one end the outer horizontal beam extends over an open-air swimming pool whose shell is

---

Lune de Sang 1号棚与2号棚 Lune de Sang Shed 1 and 2
地点：Northern NSW, Australia / 事务所：CHROFI / 项目总监：John Choi / 项目主管：Toby Breakspear
CHROFI团队：Steven Fighera, Tai Ropiha, Jerome Cateaux, Clinton Weaver, Linda Lam, Felix Rasch / 现场管理：Tony Kenway / 结构工程公司：DW Knox & Partners / 土木、水利与电气工程：Northrop / 照明顾问：Electrolight, Architectural Lighting Design / 规划者：Planners North / 成本规划：QS Plus / 丛林防火与废水：BCA Check / 认证官：Techton / 建造方：Cedar Creek Constructions / 主管：Lyle Le Sueur / 领班：Karl Vikstrom / 石匠：Robert Hartnett and Sons / 可开启门：Steve Jones, Monarch Doors / 木匠：Men Joinery / 客户：Andy & Deirdre Plummer / 功能：Storage & Workshop / 面积：Shed 1,320m², Shed 2,360m² / 竣工时间：2013 / 摄影师：©Brett Boardman (courtesy of the architect)

Lune de Sang展馆 Lune de Sang Pavilion
地点：Northern NSW, Australia / 事务所：CHROFI / CHROFI团队：John Choi, Jerome Cateaux, Eoin Healy, Clinton Weaver / 功能：Residential / 用地面积：995m² / 总楼面积：280m² / 建筑规模：one story / 竣工时间：2017 / 摄影师：©Brett Boardman (courtesy of the architect)

Lune de Sang石屋 Lune de Sang Stone House
地点：Northern NSW, Australia / 事务所：CHROFI / CHROFI团队：John Choi, Jerome Cateaux, Fraser Mudge / 团队：DW Knox & Partners, Kate Singleton, QS Plus / 建造方：Lyle Le Sueur / 客户：Private / 功能：Alterations & Additions / 总楼面积：165m² / 竣工时间：2013 / 摄影师：©Brett Boardman (courtesy of the architect)

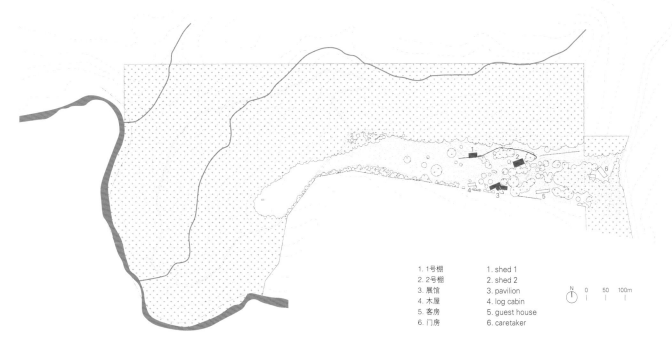

1. 1号棚	1. shed 1
2. 2号棚	2. shed 2
3. 展馆	3. pavilion
4. 木屋	4. log cabin
5. 客房	5. guest house
6. 门房	6. caretaker

N 0 50 100m

within stone wall where the land falls away.

Physically the building is one structure but to enter the private retreat from the communal space, one must step outside and pass through a door in the retaining wall that divides the two parts. The retreat's full-height glass perimeter, also set in its concrete frame, makes this a glass house. Rather than sweeping views of the landscape, it looks onto a lush rainforest embankment, lending the space a subterranean quality. Joinery of blackbutt (a eucalyptus species) timber partitions the space. The roof is defined by two rectangles which are joined and extended from the retaining wall at different angles. The bedroom and bathroom are continuous under the longer rectangular roof. The proportions and scale of the retreat are intimate and reveal the refined craftsmanship which ensures this building should last for centuries.

The Pavilion is built sustainably. The concrete and stone structure provides thermal mass that keeps the space cool. Grey water is returned back into the forest ecosystem after use. Care was taken in ensuring a minimized carbon cost and this was bolstered by using locally sourced and recycled materials. For example, the stone walls are built with rock from the forest site. Lune de Sang's two sheds were completed before the Pavilion. They are grounded, rhythmical buildings with a crisp repetition of elements that straddle earth berms and stone retaining

• 1号棚 **Shed 1**

walls. Both emerge from hillside, and the thermal mass of their concrete and earth makes the interiors cooler in summer and warmer in winter.

Shed 1, which is 320m², is formed by parallel concrete elements extending from a stone wall embankment and carrying strip lighting. They become buttresses as they fold down into 19 slab columns on its other side. The calm, sheltered volume of the shed is enclosed on three sides by glass inside of the three columns at either end.

Behind the main space, a workshop space extends into the hillside. The structure frames views into the forest in front through the rhythm of the columns, and a visual release upwards and outwards to the open paddock and fig tree, and the stone wall as it winds into the building.

Shed 2, an open structure of 360m², is the conceptual inverse of Shed 1. Its soaring zinc roof provides cover for storage. Eleven parallel 20m-long concrete roof beams emerge from the hillside to anchor the 11m partially cantilevered roof. Beneath the roof, wooden wall sections enclosing a rectangular space swing up at the front and open like doors at the sides, to facilitate materials handling and storage. When shut, the space is still open to air as the walls do not reach the roof.

In the fullness of time, both shed structures will only be revealed upon arrival at their immediate setting, intensifying their presence in the landscape.

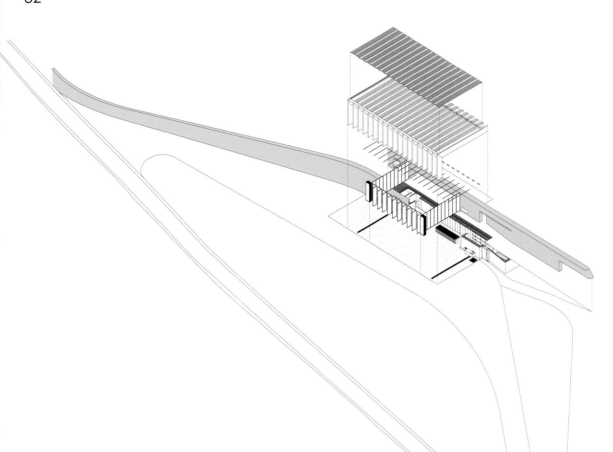

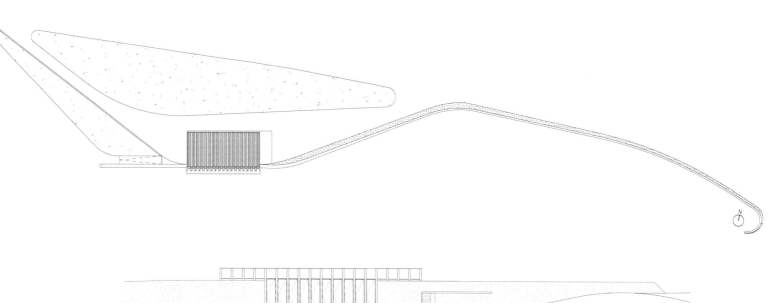

北立面 north elevation

A-A' 剖面图 section A-A'

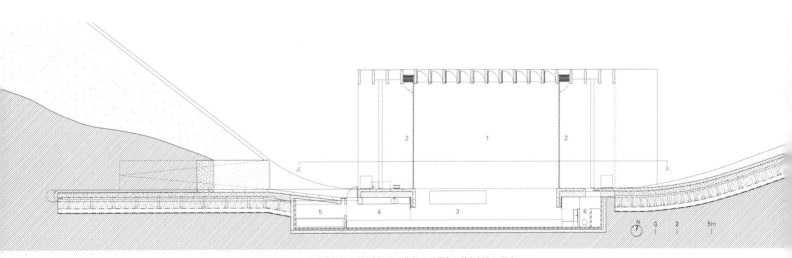

1. 工作棚区 2. 可开启门 3. 工作室 4. 小厨房 5. 储存空间 6. 浴室
1. shed area 2. operable doors 3. workshop 4. kitchenette 5. storage 6. bathroom
一层 ground floor

2号棚 Shed 2

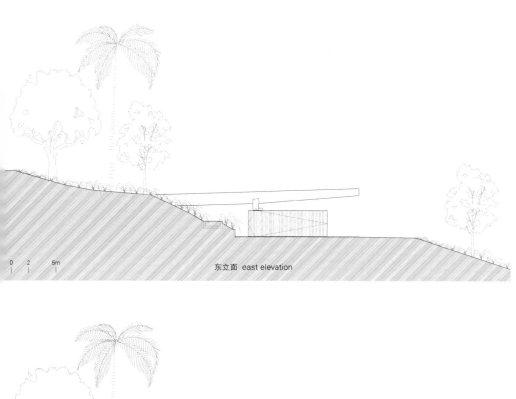
东立面 east elevation

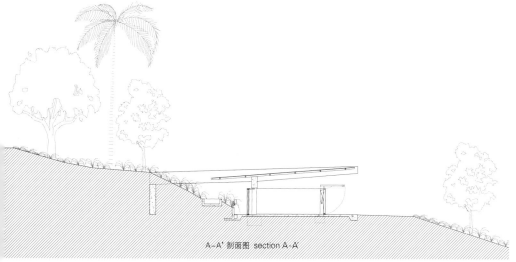
A-A' 剖面图 section A-A'

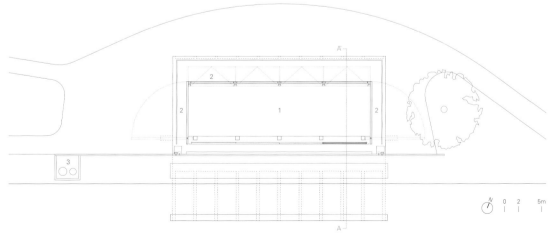

1. 工作棚区 2. 可开启门 3. 燃料室　1. shed area　2. operable doors　3. fuel stand
一层　ground floor

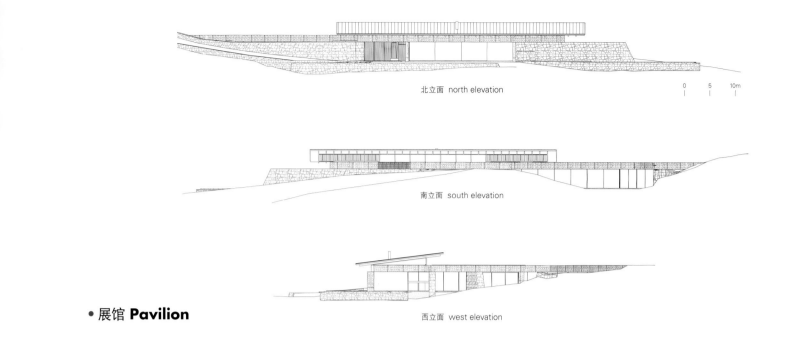

北立面 north elevation

南立面 south elevation

西立面 west elevation

0 5 10m

• 展馆 Pavilion

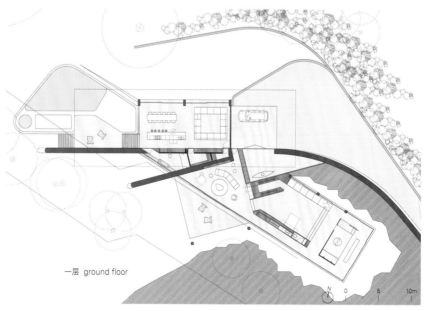

一层 ground floor

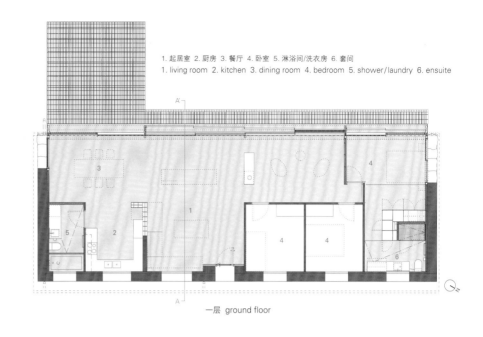

1. 起居室 2. 厨房 3. 餐厅 4. 卧室 5. 淋浴间/洗衣房 6. 套间
1. living room 2. kitchen 3. dining room 4. bedroom 5. shower/laundry 6. ensuite

一层 ground floor

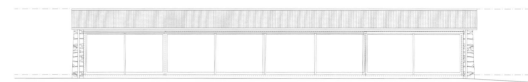

西立面 west elevation

• 石屋 **Stone House**

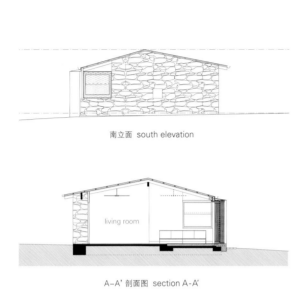

南立面 south elevation

A-A' 剖面图 section A-A'

山海美术馆
Mountain & Sea Art Museum

gad

浮于石林之上：以综合体形态解构美术馆概念
Floating above a stone forest, a complex angular form deconstructs the idea of a museum

喀斯特岩锥形柱分布于中国南方的云南省，也是翠峰生态公园山海美术馆的特色。山指的是博物馆依靠的山，海指的是博物馆俯瞰的海，但是这个海不是真正的海，而是滇池。滇池北部正是昆明市的建市核心。棱角分明的美术馆占地2998m²，与石林融为一体，石林甚至进入博物馆内部布局，整座建筑好似漂浮在山坡的树木之间。

美术馆的形态设计受到莫比乌斯带的启发。莫比乌斯带是一个环，但经过扭曲之后，外人看起来它只有一个面。为此，博物馆采用了棱角分明的建筑外形，在玻璃墙的内外都采用了垂直混合设计，另外其有机的分层表面还采用了夯土材质。在风格和形式上，美术馆都称得上是解构主义作品。每一层都是不同的，但在立面上又形成了并置效果。内部空间主要是连贯的梯形，这个梯形在宽度和高度上会有变化。从上面和内部交通路线上看，这个结构就像一个由直线构成的数字"8"。

美术馆体验始于山坡上的开放"空中平台"。这里四周围着石柱墙，可以欣赏到壮观的景色，除此之外还有为沉思者设计的专座。在岩石旁，楼梯从平台上通向美术馆建筑三层中最下面的一层。这一层如同水平悬臂向外伸出，并围绕着"8"字围合的空隙的其中一折叠。"8"字有两个空隙，中间都有树木生长。从高层向下看，可以看见低层的黑色屋顶，反射着满天繁星。在内部，展览空间始于长画廊。画廊通往进入主楼层的楼梯，而楼梯与下面的地板交错排列，围住了另一个"8"字形空隙。这层楼设有画廊、卫生间和办公室。在最大的双层画廊内，一根倾斜的白色结构柱拔地而起，并像树分叉一样分成三部分，支撑着倾斜的屋顶。最高层容纳了更多的办公室，悬于树形柱子旁的画廊空间中。

除了中国gad公司上海工作室构建的整体建筑，IDM Studio的室内装饰和种地景观设计公司的外部美化也为山海美术馆的非凡卓越立下了汗马功劳。与建筑相邻的还有一条空中步道，表面为黑色，它高耸于美术馆附近的石林上，人们可在此欣赏高处景观和地形。

美术馆的空间、角度和材料搭配与其所处位置特性密切相关。美术馆建筑形态与石林分布相吻合，使得整座建筑仿佛穿行于石林之间。在屋顶反光部分的作用下，山下湖水仿佛被舀到了建筑物上。当地的土地实际上也融入了建筑物。设计布局通透优美，将山脉、石林和植被引入其中。在建筑高处还能看到昆明市中心，美术馆本身也成了湖景的一部分。总的来说，山海美术馆是一个安静的地方，远离了城市的喧嚣。

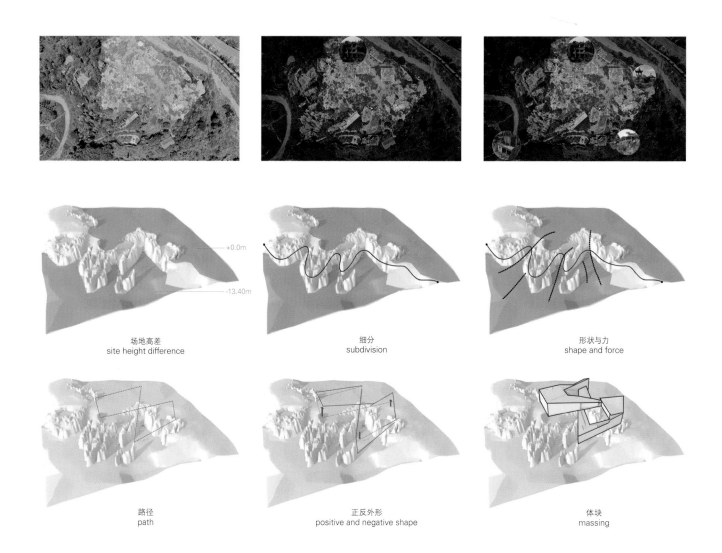

Tapering columns of karst rock outcrops in Yunnan, southern China, characterize the site of the Mountain & Sea Art Museum in CuiFeng Eco-park. The building's name refers to the mountain beyond it and the water below it, although the water is not sea but Lake Dianchi, where the city of Kunming has built up around its northern end. The angular 2,998m² museum is integrated with the stone forest, which even enters into the space of the structure, and it floats amongst the trees growing on the hillside location.

The form of the museum is inspired by a möbius strip, a loop twisted to have one side. This is manifested by an angular building with a vertical mixture both inside and outside of glass curtain walls and organically stratified surfaces of rammed earth. In style and form, the museum is deconstructivist. Every level is different, producing juxtapositions in elevation, and internal spaces are mainly connected trapezoids, changing in width and height. From above, and in its internal

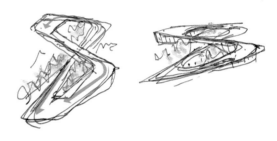

roof hanging system
suspender support mezzanine space
tree column for support

1. 门厅 2. 展览大厅 3. 游泳池 4. 办公室 5. 卫生间 6. 石林
1. lobby 2. exhibition hall 3. pool 4. office 5. rest room 6. stone forest
三层 second floor

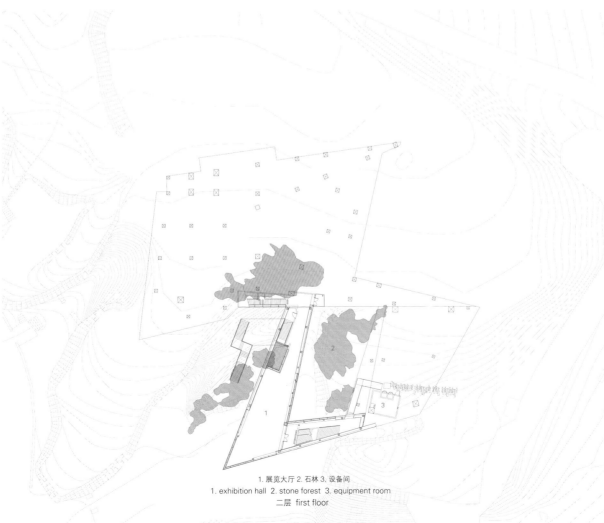

1. 展览大厅 2. 石林 3. 设备间
1. exhibition hall 2. stone forest 3. equipment room
二层 first floor

circulation route, the structure is like a figure "8" but with straight lines rather than curves.

The museum experience begins on an open "sky platform" on the hillside, bounded by walls of rock columns, which offers spectacular views and seating for contemplation. Stairs rise beside rocks from the platform to the lowest of the three levels in the building. This level cantilevers out and folds around one of the two voids enclosed by the 8 shape. Trees rise through both voids. This lower level has a black roof that reflects the sky when looked down from higher levels. Inside, the exhibition spaces begin with a long gallery that leads to stairs accessing the main floor, which is staggered from the floor below to encompass the other void. This floor contains galleries, toilets and offices. Inside the largest, double height gallery, an inclined white structural column rises and branches like a tree into three parts, supporting a sloping roof. The highest floor hosts more offices, cantilevered into the gallery space beside the tree-like column.

As well as the architecture by the Shanghai studio of Chinese practice gad, the interior decor by IDM Studio and the external landscaping by ZhongDi Landscape Design are integral to building the extraordinary impression that the Mountain & Sea Art Museum makes. There is also an adjacent skywalk surfaced in black, which rises over the stone forest adjacent to the museum, offering elevated views of it and the terrain.
The composition of the museum's spaces, angles and materials is intimately connected to the elements of its location. Its form is threaded through the stone forest. The reflective part of the roof is as if a part of the lake's water below has been scooped up onto the building. The local earth is physically part of the building. The mountain as well as the stone forest and vegetation are brought into the building by its transparency. The high position also connects with the vast metropolis of Kunming, in the vista towards the lake that it commands. But the museum is a place of tranquil spectacle, high above the city's hustle and bustle.

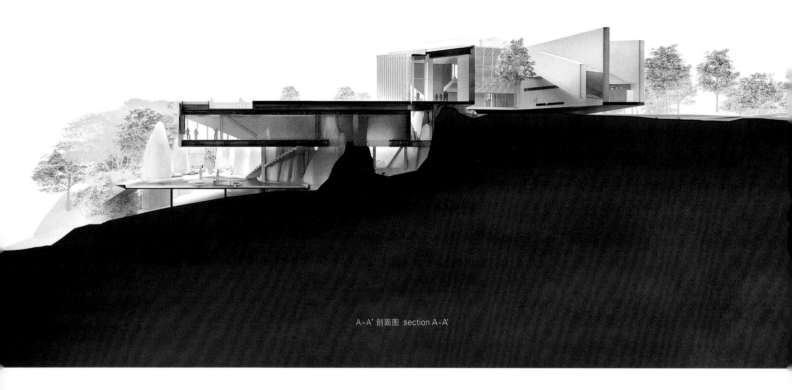

A-A' 剖面图 section A-A'

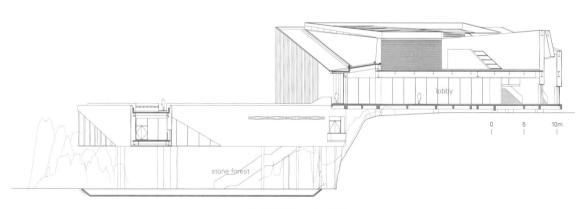

B-B' 剖面图 section B-B'

项目名称：Mountain & Sea Art Museum / 地点：Kunming City, Yunnan Province, Chi / 事务所：gad / 项目主管：Jiangfeng Wang / 项目创意：Jiangfeng Wang, Jiaying Chen
项目团队：Jiangfeng Wang, Wei Wang, Jiaying Chen, Chao Zhang, Yinghui Deng, Xin Li, Ziyi Yan, Sheng Sun, Hui Xu / 结构：Jie Wang, Yuan Ren, Liang Huang, Long Zhang, Zhan Shi, Zhuo Chen, Yichao Xi / 给排水：Ning Tuo, Mengxiao Wang, Hao Zheng, Hai Sun / 暖通空调：Dake Mao, Zhifeng Xu, Yafeng Wei
电气工程：Jin Wang, Xiaoyun Chen, Feng Sun, Jianpeng Zhang, Weijie Yan / 景观设计：ZhongDi Landscape Design Co., Ltd. (Beijing, Shanghai) / 室内设计：IDMatrix
软装设计：Matrix Mingcui / 建筑所有者：Yango Group Co., Ltd. / 项目类型：Cultural Display / 总楼面面积：2,997.88m² / 设计时间：2019.7 / 竣工时间：2020.4
摄影师：©Guangkun Yang (courtesy of gad) - p.104~105, p.106[top], p.115, p.117; ©Shenzhen Shi Xiang Wan He Cultural Communication Ltd. (courtesy of IDMatrix) - p.100~101, p.103[upper], p.108, p.109, p.110~111, p.112~113

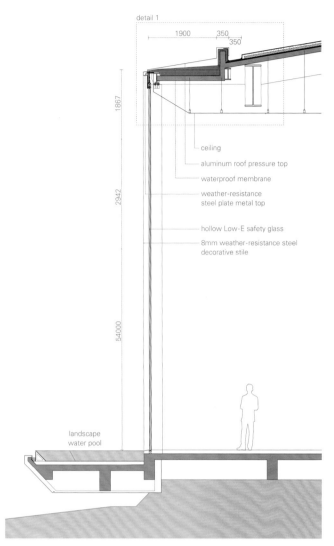

a-a' 详图 detail a-a'

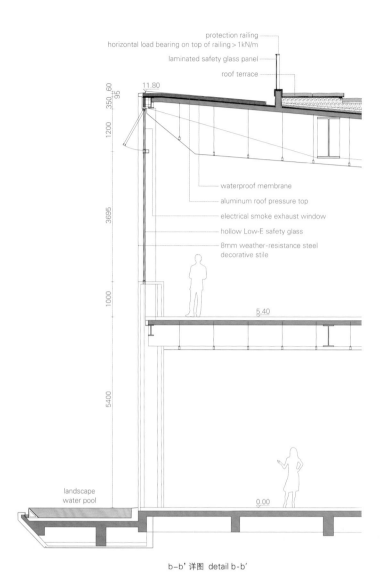

b-b' 详图 detail b-b'

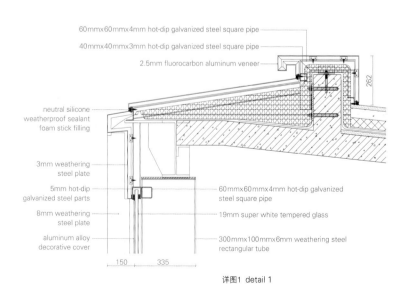

详图1 detail 1

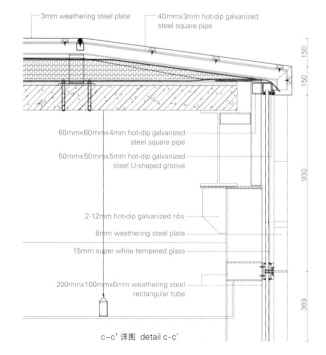

c-c' 详图 detail c-c'

周口店北京猿人洞1号考古遗址保护棚
The Protective Shelter of
Locality 1 Archaeological Site of Zhoukoudian Peking Man Cave

THAD

棚罩与景观融为一体，悬于建筑之上，为北京猿人洞遗址及其相关考古成果提供保护
A floating canopy protects a Chinese cave and its archeology and merges with the landscape

周口店位于北京西南42km处，那里有着被列入联合国教科文组织世界文化遗产的北京人遗址。这片遗址位于树木繁茂的山坡上，曾是50万年前的史前人类居住地；现在，清华大学建筑设计研究院为其设计了悬浮棚罩，以提供保护。

遗址受到侵蚀、风化，面临破坏和不稳定的威胁。在2012年的暴雨之后，洞穴中出现了积水。通过物探，人们在洞穴中发现了大面积裂缝和断裂带，危及遗址的整体稳定。2013年，国际古迹和遗址理事会（ICOMOS）批准了一项新的保护计划，设想为遗址建造一个保护棚。2014年，建筑图纸完成；2018年，建设完毕。这个保护棚很快就向公众开放了。

保护

保护棚的设计目的是保护遗址免受雨、雪、冰雹和其他威胁因素的影响。建筑采用了保护策略，减少了风、温度和湿度的剧烈波动所带来的影响，同时也保持了遗址与原始自然环境的关联。

结构

该项目应用了最小干预和可逆干预原则。两排支撑脚分别位于南面的山峰和北面的山脚，支撑着大跨度的网格状外壳，使之覆盖洞穴的核心部分，避免与遗址和悬崖接触。支撑脚位于平坦的岩层上，外壳承载着两层方形钢叶片，一层安装在钢结构的上方，另一层悬挂在钢结构的下方。棚罩所有组件都是在现场预制和组装的，以最大限度地减少施工过程对现场的干扰，并确保该顶棚在未来可被拆除，从而恢复其原有的环境条件。

同时，在北京猿人洞微环境气象监测数据的基础上，设计方利用风洞技术进行模拟计算，对遮挡高度、钢叶片间隙、温度、湿度、采光、通风等参数进行模拟。计算结果证明设计是正确的。双层表皮的设计形成了缓冲层，它可以缓解一些因素的突然变化，如温度和湿度的突变。

壳体结构不规则曲面面积3728m²，东西跨度54.5m，南北跨度77.5m。两排支撑脚的高差为30.7m，结构高度为37.4m，棚罩最大离地高度为35.7m。

叶片的结构是通过参数化计算产生的，它们紧密地相互重叠，以实现对雨水的高效排出，同时通过它们之间的间隙提供统一的采光和通风。这两层表皮共有825片叶片，其中外表皮420片，内表皮405片。

与环境融合

北京猿人洞原本是由山的自然形态塑造的，但在洞穴岩石倒塌之后，设计方通过现在的地形轮廓等探查原地形，塑造整个棚罩的形态。因此，新结构能够与周围环境的地形相融合。外部叶片显露自身的钢或铝表面，但按照设计，植被可以向下延伸到叶片上，因此外层成为一个绿色屋顶，经过植被两年生长，屋顶会逐渐隐藏在周围的树林中。与此同时，内层的叶片的下面有一个类似岩石的图案，以呼应下面的岩石洞穴环境。

游客设施

在游客设施方面，游客入口位于棚罩脚下的椭圆形屋顶结构中，一部木楼梯小路从这里穿过岩石，通向顶棚下方。在北京猿人洞的悬崖上，有沉浸式多媒体展示，向游客说明该遗址的挖掘过程、历史背景和研究成果。

A UNESCO World Heritage Site cave known as Peking Man Site, at Zhoukoudian, 42km southwest of Beijing, contains relics of prehistoric human occupation that are up to half a million years old. Located in a wooded hillside, it is now protected by a floating canopy designed by THAD (Architectural Design & Research Institute of Tsinghua University).
The site has been threatened by erosion, weathering, crumbling and instability. After the severe rain storm in 2012, water catchment was discovered in the cave. Through geophysical prospecting, large fissures and fracture zones were found which endangered the overall stability of the site. A plan of a new protective shelter was approved by ICOMOS (the International Council of Monuments and Sites) in 2013, construction drawings were finished in 2014, and construction was completed in 2018. The shelter was swiftly opened to the public.

Conservation

The protective shelter is designed to protect the site from

rain, snow, hail and other threatening factors. A conservation strategy is applied so that the shelter can reduce violent fluctuations of wind, temperature and humidity, while keeping the site connected to its original natural environment.

Structure

A principle of minimal and reversible interventions is applied to the project. A grid-like shell structure of large span is supported by two rows of footholds situated on the peak of the hill on the south and at the foot on the north, to cover the minimum necessary area of the cave and to avoid contact with the site and its precipice. The footholds are situated on the flat rock formation. The shell carries two layers of square steel vanes or blades, one mounted above the steel carrying structure and the other suspended underneath it. All the components of the shelter are prefabricated and assembled on site to minimize disturbance on the site during construction work, and to insure the possibility of dismantling in

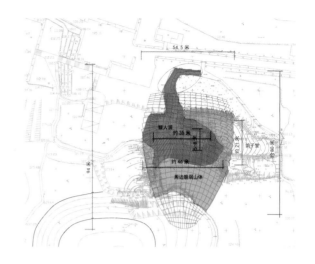

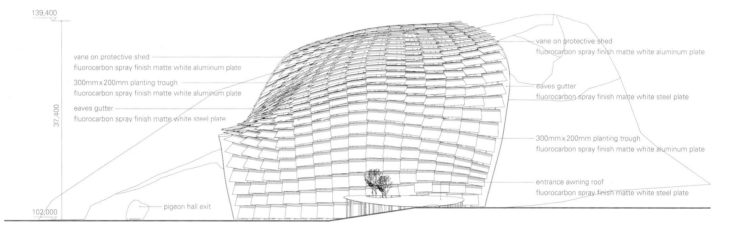

南立面 south elevation

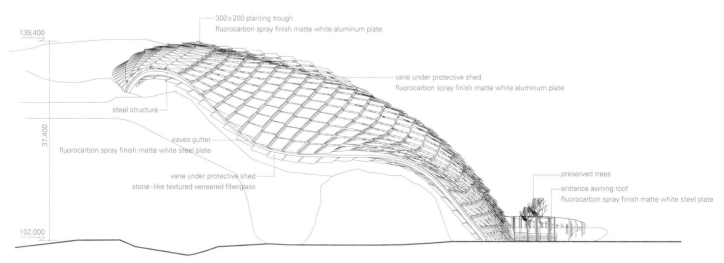

西立面 west elevation

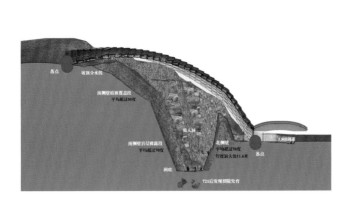

future to restore the environment to its original condition. Based on meteorological monitoring data of the micro-environment in Peking Man Cave, analog computations conducted with wind tunnel techniques are conducted to simulate the height of the shelter, gaps between blades, temperature, humidity, daylight, ventilation and other parameters. The results vindicate the design. The double-skin creates a buffering layer to alleviate abrupt changes such as in temperature and humidity.

The irregular curved surface of the shell structure covers an area of 3,728m². The east-west span is 54.5m and the north-south 77.5m. The height difference between the two rows of footholds is 30.7m, the height of the structure is 37.4m and the maximum height of the shelter from the ground outside is 35.7m.

The configuration of the blades is generated by parametric calculation, and they overlap closely with one another to enable efficient rainwater drainage, while providing uniform daylight and ventilation through the gaps between them. There are 825 blades in all, of which 420 blades are on the outer skin and 405 on the inter skin.

Blending with the environment

The morphology of the hill before the rock collapse that created the Peking Man Cave was deduced from clues such as contour lines of the current landform, to shape the overall form of the shelter. The new structure is consequently able to blend with the topography of its immediate surroundings. The outer blades expose steel or aluminum faces but are designed to allow vegetation to extend downwards over them, so that the outer layer becomes a green roof which gradually hides itself into the surrounding woods after two years of growth. Meanwhile, the underside of the inner skin's blades has a rock-like pattern, in response to the rocky cave environment underneath.

Visitor facilities

The visitor entrance is in an oval-roofed structure at the foot of the shelter, and from it a wooden path with stairs leads under the shelter and through the rock. The precipice of Peking Man Cave hosts an immersive multimedia display explaining the excavation process and historical backgrounds and research findings.

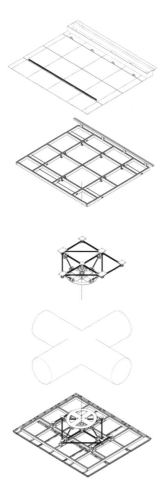

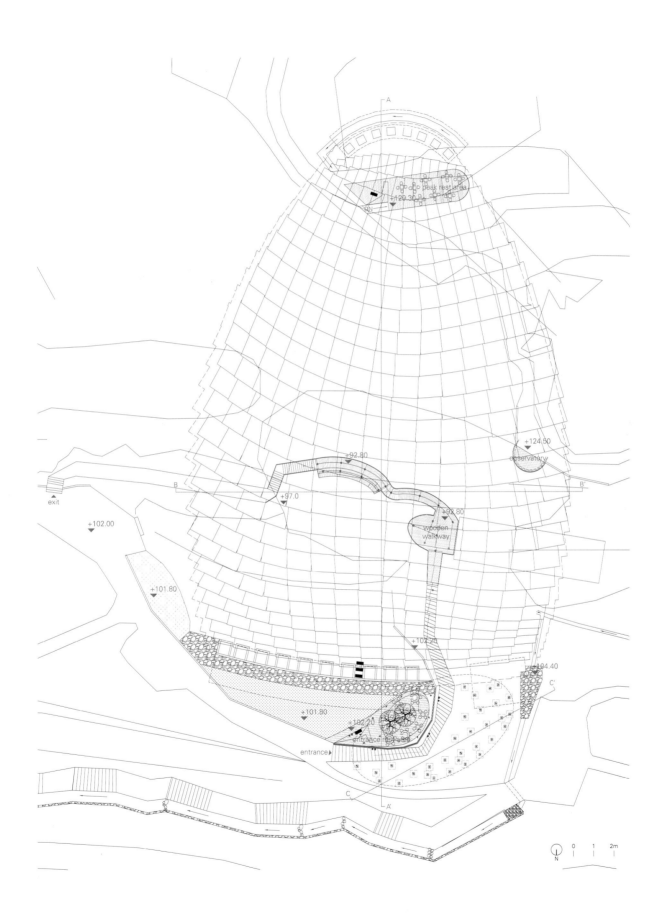

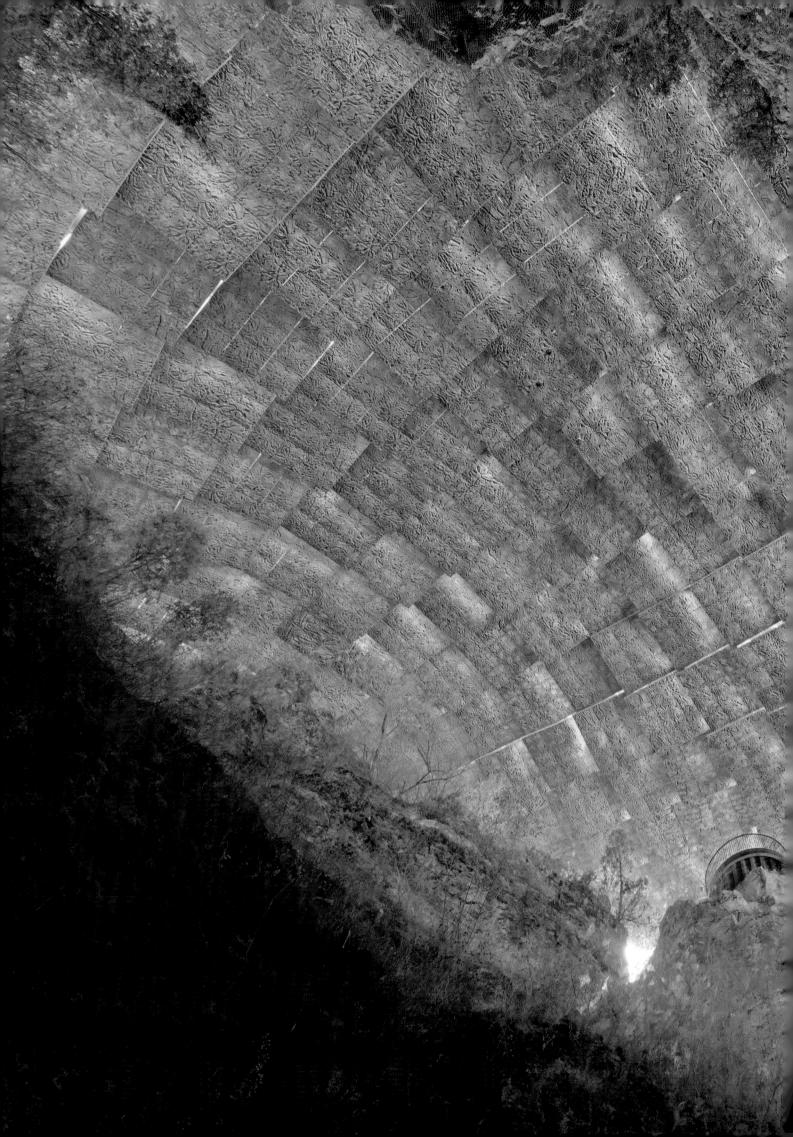

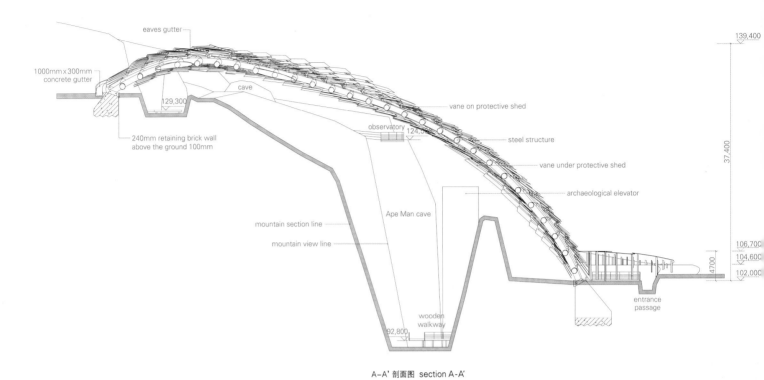

A-A' 剖面图 section A-A'

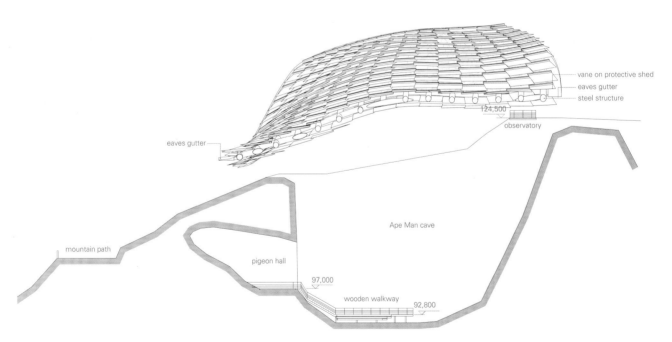

B-B' 剖面图 section B-B'

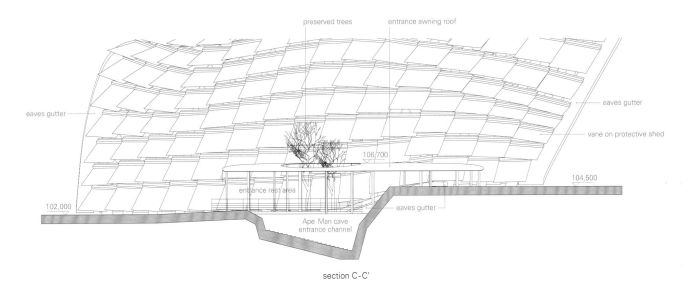

section C-C'

项目名称：The Protective Shelter of Locality 1 Archaeological Site of Zhoukoudian Peking Man Cave / 地点：China, Beijing / 事务所：Architectural Design & Research Institute of Tsinghua University (THAD) / 设计团队：general leader – Lu Zhou; chief architect – Cui Guanghai; architects – Wang Jing, Jie Xiaofeng, Li Jing / 结构工程：Ma Zhigang, Li Zengchao, Jiang Bingli / 消防与管道工程：Guo Hanying, Liu Jiuling, Wang Jing / 电气工程：Guo Hongyan, Zhang Wei, Wang Haotian / WHC-UNESCO报告员：Xu Zhilan, Luo Xuan / 承包商：Beijing Jianlong Heavy Industry Group Co., Ltd / 项目经理：Yuan Jixiang / 客户：Zhoukoudian Site Museum / 用途：heritage conservation / 用地面积：2,878m² / 总楼面面积：3,487m² / 竣工时间：2018 / 摄影师：courtesy of the architect (except as noted)

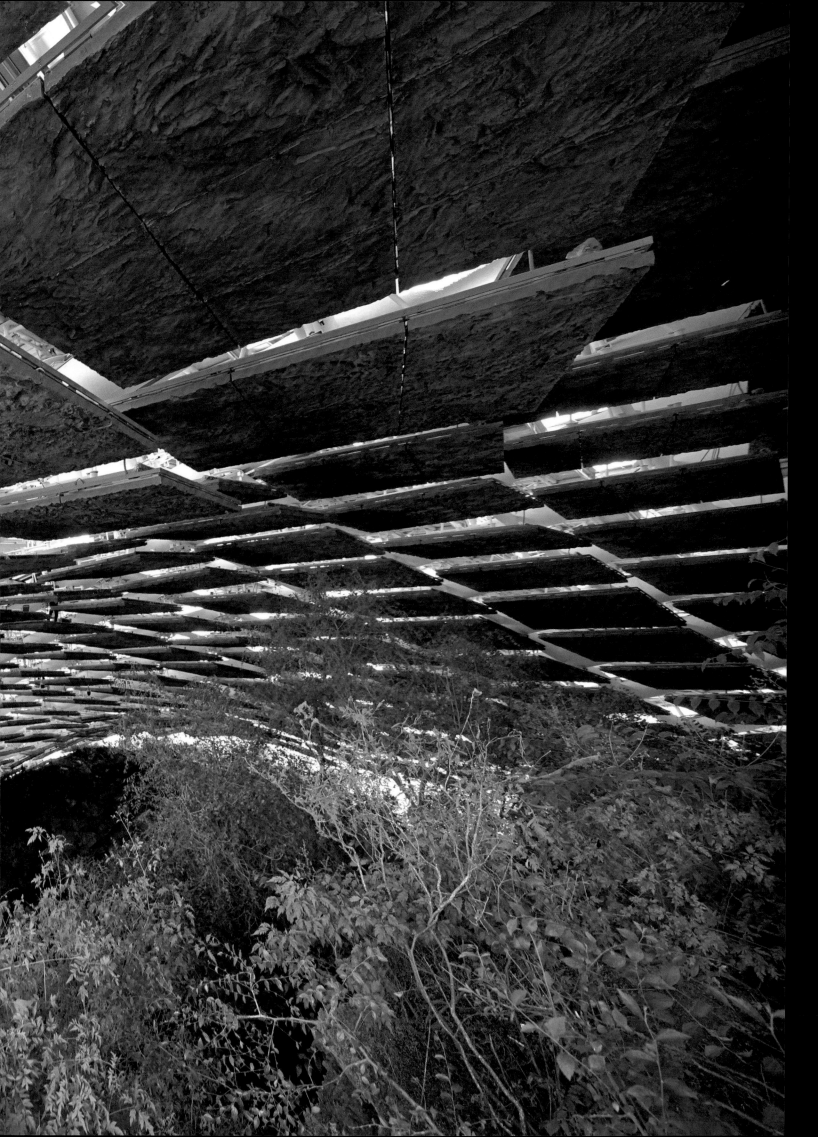

艺术生物栖息地水景园
Art Biotop Water Garden
Junya.Ishigami + Associates

池塘群中重植树木，打造全新类自然人工环境
By replanting trees in a network of ponds, a new nature-like artificial environment is created

本案例位于日本栃木县。建筑师石上纯也在艺术生物岛设计了一处植物园，并将其打造为艺术休养地。除此之外，项目所处的草地经过设置，成为由弧形的池塘组成的区域。就在这片土地上，设计师重新种植了一片森林，打造了宁静而引人沉思的全新环境。

建造宾馆需要清空森林区域。于是设计师决定将森林重新移植到附近一块较小的场地上。该处面积16 670m²，本是一片稻田，后来变成了一片草地。四年间，设计师对318棵树（包括金合欢、青木、山毛榉、枫树和橡树）进行了测量和单独编号，并将每棵树重新定位到指定的新种植点。树木都被安置在160个新设计的不同形状的池塘中，有些池塘中还裸露着很多小片土地。种植过程中使用了传统的工具来调整水平面。落叶树种会因为靠近水体而淹死，但设计师在农田土壤下方铺设了膨润土薄膜，防止池塘水随意外溢。现场原有的水闸通过管道将水注入池塘，使得水以不同的流速流动。该地区周围的苔藓被重新种植在这里。场地上的石头也被重新安置，以创造时不时可以穿过水面的路径。

因此，水景园将长满苔藓的森林叠加在稻田景观上，保留了场地的历史痕迹。设计师通过规划特定的树木设计池塘形状，为朦胧的森林景观添加框架，创造了具有众多细节的空间。在这一景观中，树木和池塘的密度，树木和池塘的共存，都不会发生在自然界中，但在这里却通过设计实现了。随着时间的流逝，森林也将随着季节的变化而生长变化。

人工自然环境是一个植根于日本园林历史的概念。建筑师指出，通过对自然的细致规划，自然环境与人工环境更加融合、融汇、融洽。

In the Japanese prefecture of Tochigi, architect Junya Ishigami has created a botanical landscape garden at Art Biotop, an art retreat. A forest has been replanted in a meadow which has been engineered to become a field of curving ponds, creating a completely new environment that is tranquil and contemplative. The development of a hotel required the clearance of a forest area. It was decided to relocate the trees to an adjacent, slightly smaller site of 16,670m², which had been a paddy field and subsequently become a meadow. Over four years, 318 trees were measured and individually numbered, and each was relocated to a specific new point. They include species of acacia, aohada, beech, maple and oak. Each tree is sited amongst the 160 new ponds which have been designed in different shapes, some with multiple lobes and islands. Traditional tools are used to level the water. Deciduous trees would be drowned by proximity to bodies of water, but the ponds are waterproofed with bentonite sheeting, which is covered with field soil. A pre-existing sluice gate on the site brings water to fill the ponds, which are connected with pipes, enabling flow at different rates. Moss from around the area is replanted. Stones on the site are also relocated to create paths that sometimes cross water.

The water garden thus superimposes the mossy forest onto the landscapes of the paddy field, retaining traces of the site's history. By planning specific shapes of trees and ponds, the vague scenery of the forest is given a framework, and considered as a space with as much detail as possible. The density of trees and ponds, and their co-existence, would not occur in nature, but here are created by design. The passage of time is built in as the rearranged forest grows and changes with the seasons.

An environment of artificial nature is a concept that is rooted in the history of Japanese gardens. The architect states that by planning nature in a detailed way in which natural environment and human environment mingle, intertwine and merge more closely.

项目名称：Botanical Garden Art Biotop/Water Garden / 地点：Tochigi, Japan / 事务所：Junya.Ishigami + Associates / 项目建筑师：Junya Ishigami, Eiko Tomura, Taeko Abe, Lucie Loosen, Gaku Inoue, Akira Uchimura, Masayuki Asami / 承包商：Shizuoka Green Service Co. / 功能：garden / 总用地面积：16,670m² / 施工时间：2013—2018 / 摄影师：courtesy of nikissimo Inc.-p.136~137, p.142~143[left]; courtesy of the architect-p.143[right-top, right-bottom], p.146~147

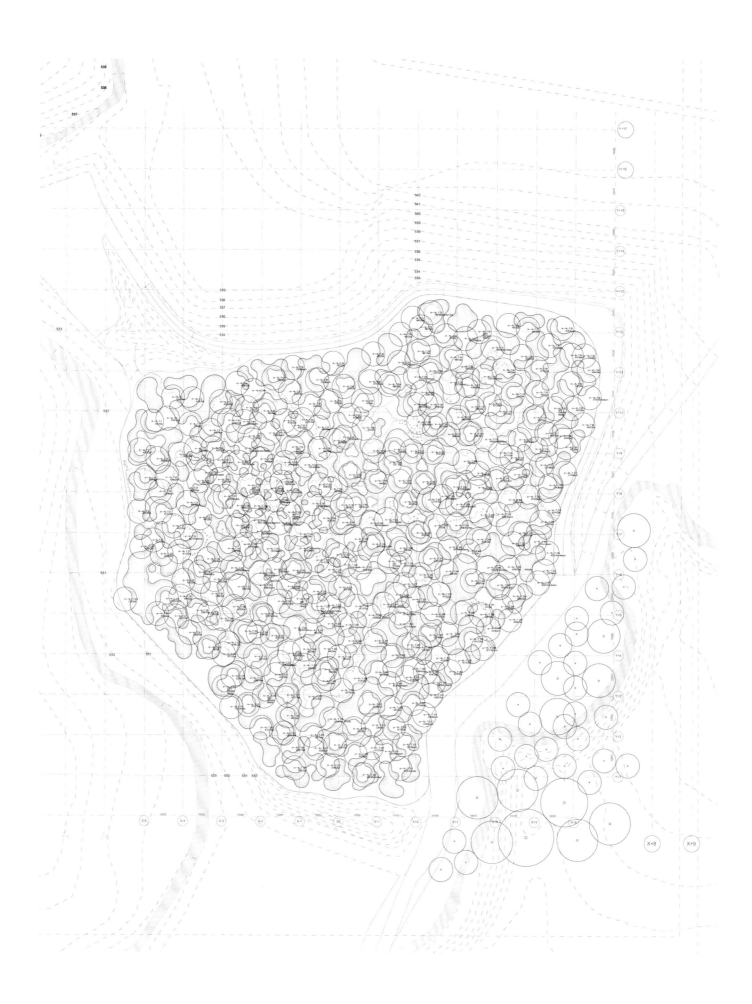

141

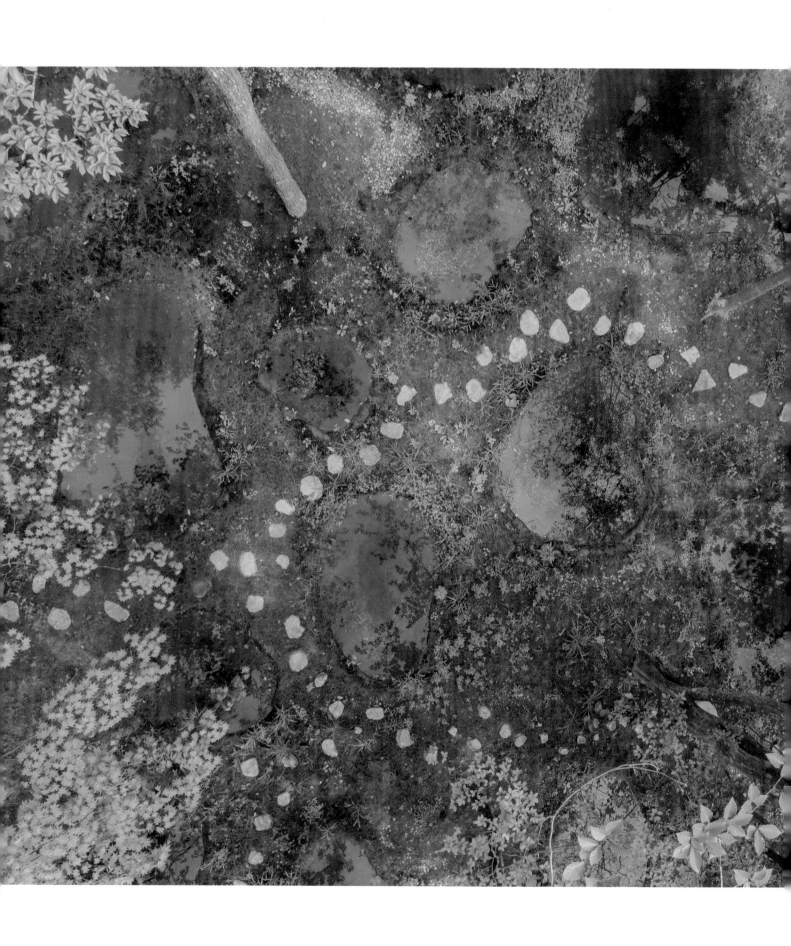

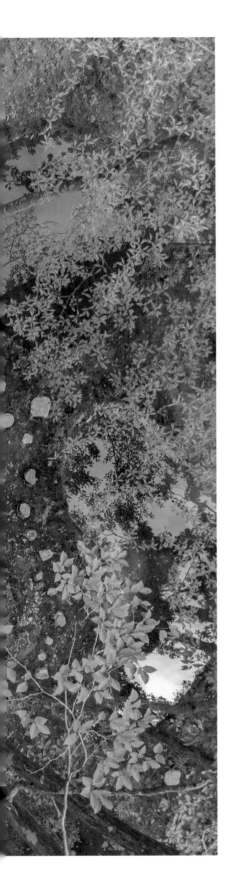

X-1剖面图 X-1 Section

社区市场：文明的纽带

Community
Links in Civili

地方市场大厅是重要的建筑类型。它不仅是食品生产者和消费者的交易平台，还可以作为社区服务设施，在塑造社区方面发挥相应的作用。除此之外，地方市场大厅还是城乡联系的载体。然而，零售商场兴起，生产工业化，交通运输发展，城市延伸扩张，法人企业超越个体户不断壮大……凡此种种，都对市场大厅的功能构成了挑战。更有甚者，绅士化、旅游业、卫生问题以及数字化等当代新生事物，更为市场大厅带来了空前严峻的压力。

A local market hall is a key architectural typology. Its function extends beyond merely offering a forum in which food producers and consumers can trade. It can be considered as a community service, playing a role in defining communities. Third, it embodies the connection between city and countryside. The roles of market halls have been challenged by the rise of retail premises, industrialisation of production, transport development, urban sprawl and the rise of corporate enterprise at the expense of independent businesses. Contemporary phenomena which add urgent pressure to market halls include gentrification, tourism, hygiene issues and digitalisation.

阿泽尔格河畔拉米尔的有顶市场_The Covered Market of Lamure-sur-Azergues / Elisabeth Polzella Architecte
格拉莫罗特市场广场_Gramalote Market Square/ Niro Arquitectura + OAU
圣雷莫安诺纳里奥市场重建_Restructuring of the Sanremo Annonario Market / Calvi Ceschia Viganò Architetti Associati
胜利市场临时安置点_Temporary Site of Shengli Market / LUO Studio
纳吉克罗斯市场大厅_Market Hall of Nagykőrös / Kiss-Járomi Architect Studio
特拉斯科工匠市场_Tlaxco Artisans Market / Vrtical

社区市场：文明的纽带_Community Markets: Links in Civilization / Herbert Wright

建筑正是市场大厅化解压力的利器。市场大厅的变化缓慢，不过建筑风格却变化频繁。据我们调查，旧市场经常通过建筑实现复苏。

无论市场大厅是遗迹还是新建筑，建筑师们都必须考虑实际问题，例如，健康的地方市场需要什么样的建筑结构？市场如何才能体现地方特色？此次调研会给您答案。除此之外，调查还包含某些涉及人际互动、城市社区以及地球生态系统的深层次问题。

Architecture is a tool with which market halls can fight back. Although architectural styles have come and gone, market halls have changed slowly. As our survey will reveal, the application of architecture is often in reviving old markets.
Whether a market hall is a heritage structure or new building, architects have to engage with practical issues, such as what does a healthy local market need that a physical structure can deliver? How can a market assert local identity? The answers emerge in our survey. Beyond the answers lie deeper questions that touch on human interaction, urban communities, and the planetary ecosystem.

社区市场：文明的纽带
Community Markets: Links in Civilization

Herbert Wright

　　远在市场建筑出现之前，就有交易商品的地方，它是文明的基石。农业生产出现过剩后，人类开始进行城市化改造，职业分化为农民、手工业者、管理者等，市场由此产生。它为社区提供食物和其他商品，成为城市人口生存的中心。今天，许多国家的乡村和城市已经成为彼此隔绝的两个社会，而市场作为农产品的直销地，依然维持着城乡纽带的功能。凡此种种，使得市场建筑成为重要的建筑类型。

　　长期以来，从仓库铁路到当代数字化配送服务，物流和运输的创新削弱了农村供应方和城市消费者之间的重要联系。食品加工业的出现改变了一切。现在，食品饮料已是全球消费市场的一部分。新鲜产品变为商品，经过加工处理、改造包装后，再被运往世界各地。我们的商品和日常生活不再朴实，推动这一变化的关键就在于市场营销和品牌宣传。市场大厅出现了用途转换趋势，例如，逐渐变为高档餐厅，以迎合绅士化浪潮。

　　但值得庆幸的是，有迹象表明社区市场正在复苏。为更好地了解社区市场，我们首先从它鼎鼎大名的古代原型入手。在古希腊，广场位于城市中心，为人们的交流会面提供了便利的公共空间，而公共市场区更是其中十分活跃的区域。人工建成的市场可以追溯至商人在广场中倚仗的柱廊，很久之后才出现专为容纳市场而建的建筑物。公元前436年，雅典帕特农神庙建成，它是最具标志性的陶立克式建筑，俯瞰着雅典广场。建筑历史学家Goerd Peschken提出，陶立克式神庙起源于古代粮仓，神庙的柱子对应的是将粮仓抬离地面的柱子。无论真假，陶立克式神庙都会让人想起城乡间无处不在的联系。

　　此次调研涉及的六个市场大厅并非购物商场、百货商店或购物中心，而是初级生产者可以直接向消费者销售产品的市场，其中几处也可被称为农贸市场。

Even before buildings emerged to contain the marketplace, a place to trade commodities was a cornerstone of civilization. It arose after agriculture created surpluses, humanity started to urbanize, and people's work differentiated, for example between farmers, craftsmen or administrators. A location to supply communities with food and other goods made the marketplace central to an urban population's survival. As an outlet for agricultural products, the marketplace still sustains links between country and city, which in many countries have become almost separate societies. All this makes the market building a crucial typology in the built environment.

The essential link between rural suppliers and urban consumers has long been eroded by the intervention of logistics and transport innovations, from warehouses and railways to contemporary digitally-linked delivery services. The emergence of industrial food processing changed everything. Food and drink is now part of the global consumer market, in which fresh products are commoditized, treated, transformed, packaged and transported anywhere and everywhere. Marketing and branding are at the heart of a demise of authenticity of what we buy, and in our day-to-day lives. An emerging trend is the conversion of market halls to other uses such as upmarket catering, a response to gentrification.

Happily, there are signs that a revival of community markets is underway. To understand them better, let us first consider a highly significant ancestor. The public marketplace was an active part of Ancient Greek agoras, the civic spaces in urban centers where the population could meet and interact. We can trace the built marketplace to when traders established themselves in colonnades in the agora, long before the buildings built specifically to shelter market places which followed. When the Parthenon in Athens, the most iconic of Doric temples, was completed in 436 BC, it overlooked Athen's original agora. The architectural historian Goerd Peschken has proposed that the form of Doric temples was descended from ancient granary stores, their columns corresponding to those lifting the granary volume off the ground. True or not, it brings to mind the ever-present connection between countryside and city.

The six market halls we survey are not shopping arcades, department stores or shopping centers, but markets where primary producers can sell directly to consumers. Some could be called farmer's markets.

中国濮阳胜利市场临时安置点（186页）由罗宇杰工作室设计。该市场本质上是极简风格的玻璃大棚，在显眼的指示牌下是摊贩一排排的摊位。玻璃大棚派生于约瑟夫·帕克斯顿的伦敦水晶宫（1851年）。水晶宫由预制部件组装而成，覆盖区域非常广阔，获取的光照度最佳。新建成的胜利市场与此完全一致，同时建筑师也处理了关于旧胜利市场等传统市场的"视觉混乱"问题。新市场布局极其开放，采用了倒金字塔框架式指示牌结构，在不影响视线的情况下能让人看到下层出售的农产品，给人以一种清新明晰的整体印象，精准满足消费者购买时的需求。

纳吉克罗斯新市场大厅（196页）需要有非同凡响的观感，一是因为这座匈牙利小镇是一座历史建筑之城，对周围新建筑的要求很高，二是因为匈牙利是一个超凡市场大厅之国。奇斯－杰罗姆建筑工作室位于匈牙利首都布达佩斯，那里有一些建筑风格迥异的市场大厅，无论是萨姆·佩茨设计的大市场（1897年），还是拉斯洛·拉伊克设计的雷赫尔市场（2002年），都给人以鲜明的印象。前者采用了宏大的新哥特外立面，后者则别出心裁地采用了狂热的折衷解构主义设计。而纳吉克罗斯市场大厅则要凸显繁荣活力。市场主外立面采用金色铝外表皮和圆形穿孔设计，入口处用大号字母拼出匈牙利语的"市场"。在令人印象深刻的正面之后，充足的自然光通过外立面和天窗洒满整个市场内部；在其他方面，市场显得较为传统。建筑师们的灵感来源于作为人们聚集空间的古希腊广场，不过市场外面的聚集空间更大，还能欣赏到全部的外立面，给人一种视觉享受。

意大利圣雷莫的安诺纳里奥市场于1960年竣工，采用了高抛物线形弯曲屋顶设计，堪称战后意大利混凝土技术的典范。然而，随着物流服务的兴起，它已大不如前。当代建筑师事务所卡尔维·塞斯基亚·维加诺修复了该市场的建筑结构，使空间布局更加清晰（176页）。虽然市场主要依靠人工照明采光，但清理挡住高层窗户的障碍后，自然光也可以照射进

The design of the temporary Shengli Market in the Chinese city of Puyang by LUO Studio (p.186) is essentially a minimalist glass shed with a field of vendor's stalls arranged in rows under highly visible signage. The glass shed is ultimately descended from Joseph Paxton's Crystal Palace (1851) in London, which was assembled from prefabricated elements to cover a large area and bring maximum light into it. This is exactly the same as the new Shengli Market. The new market also responds to a problem identified by the architects as "chaotic visual identity" in traditional markets, such as the old Shengli Market. This is addressed by the new market's exceptional openness and its signage structures, which are upside-down pyramid frames, to identify the sort of products being sold below them without interrupting clear lines of sight. They contribute to an overall impression of freshness and clarity, exactly what customers want when buying.

A new Market Hall in Nagykőrös (p.196) needed to look exceptional because the Hungarian town is full of historic buildings, demanding a lot from a new neighbor, and also because Hungary is a country of exceptional market halls. The architects Kiss-Járomi Architect Studio is based in the capital Budapest, which has several market halls in distinct architectural styles, from the Great Market Hall (1897) by Samu Pecz with its grand neo-gothic facade to Lehel Market (2002) by László Rajk, a startling, madly eclectic deconstructivist design. The Nagykörös market hall shares a visible exuberance because big letters suggesting the word "market" in Hungarian shape the entrances in its main facade, which has an outer skin of gold-colored aluminum perforated by circles. Behind this impressive front, the building is generously filled with natural light through its facades and skylights, but is otherwise more conventional. The architects were inspired by the Greek agora as a gathering space, although many will gather outside where there's more space and the sight of the full facade to enjoy.

In Sanremo, Italy, the Annonario Market was completed in 1960 and the market hall lied under a soaring parabolically curving roof, an example of post-war Italian mastery of concrete. It had fallen into decline physically and witnessed the rise of delivery services. Contemporary architects Calvi Ceschia Viganò addressed structural restoration, and has brought clarity to the space (p.176). The market relied on artificial light, but clearing obstructions to the high-level windows now allows natural light. Order came to the market floor with

来。新设计采用了整齐有序的木框架，取代了之前杂乱无章的开放摊位，构建了市场秩序。

人们对市场秩序的需求可能被夸大了，因为无论是货物运输物流，还是近距离展示琳琅满目、五花八门的商品，抑或是顾客在市场中来来往往的路线，都说明市场运营有其复杂的一面。市场也许是一台将商品从生产者转至消费者的机器，但不必像火箭发动机那样经过精心设计，专为高效运转而生，它是自组织的非正式场所，兼具社交性和娱乐性。不过，构建市场秩序的确能解决物流、交通等问题，胜利市场和安诺纳里奥市场就通过构建市场秩序，营造了健康清爽的环境。

和纳吉克罗斯一样，法国小镇阿泽尔格河畔拉米尔也建设了新市场（154页）。这是一个开放的市场大厅，庞大的屋顶雄踞其上，遮蔽了整个市场。不过，屋顶设置了侧面百叶和部分玻璃瓦，阳光得以透入市场。市场还采用了当地的木材和石料，这样不仅有助于可持续发展，也保留了当地特色。虽然建筑师伊丽莎白·波尔泽拉声称这座建筑并没有突出的风格，但鉴于它运用了历史主义风格支柱、坡屋顶，甚至散发着古希腊陶立克式神庙的气息，我们还是可以将这座市场归类为后现代主义作品。

在此次调研中，除一个市场外，其余市场均使用了木材。甚至连胜利市场在摊位和指示牌上使用钢铁和玻璃也只是为了达到较好的效果。这是因为木材是一种有机材料，它可以与食物产生一种令人安心的共鸣，并在建筑结构中构建人与自然的联结。另外一种所有这些项目都用到的材料就是自然光。它不仅让商家和顾客心情愉悦，还满足了市场有关展示新鲜农产品卫生状态的要求，让顾客能看清楚自己购买的产品。

与圣雷莫市场类似，墨西哥特拉斯科工匠市场（208页）是对一栋停业市场建筑的翻新。虽然它是本次调研中唯一不销售食品的市场，但商贩仍是市场的主要生产者，他们会向游客兜售自己制作的手工艺品。Vrtical建筑事务所为其设计了全新摊位，开放内部空间，让光线通过新天窗透入市场。事务所将新屋椽排成魅力十足的长排，整体凸显大厅长度，并在

a new design of wood-framed vendor box units lined up in neat, ordered rows, replacing the previous open, sprawling stalls.

The need for order can be overstated, because a market place operates with complexity, from the logistics of bringing in goods to the visual cacophony of having diverse, colorful products on display in proximity, and the different path every customer makes inside. A market may be a machine to transfer goods from producers to consumers, but it doesn't need to be engineered for maximum efficiency like a rocket engine. A market self-organises, retaining an informality that makes the market experience social and pleasant. Nevertheless, introducing order addresses issues such as logistics and circulation, and at the Shengli and Annario markets, it has created a healthy, airy environment.

Like Nagykőrös, the French town of Lamure-sur-Azergues has gained a new market (p.154), in this case in an open market hall under a dominant roof that is so big that its shadow can cover the market floor. Nevertheless, with louvered sides and partial glass tiling, the roof filters light beneath it. Its use of local wood and stone contributes to sustainability and asserts local identity. Although architect Elisabeth Polzella claims the structure has no style, it could be called Post-Modern with its historicist use of columns and pitched roof, even suggesting an ancient Greek Doric temple.

All our projects except one take advantage of the materiality of wood. Even the steel and glass, Shengli market uses them for good effect in its stalls and signage boards. Because wood is an organic material, there is a re-assuring resonance with food products, and it brings a connection with nature into the built structures. Another material all the projects exploit is natural light, which not only is uplifting for traders and customers, but also meets a market's need to reveal hygiene in fresh products and let customers see what they may buy clearly.

As with Sanremo's market, the Tlaxco Artisans Market in Mexico (p.208) is a renewal of an existing market building, which in this case had closed. This is the only non-food market in our survey, but the vendors are still primary producers, making handicrafts to sell to tourists. The architects Vrtical have designed completely new market stalls and opened up the space, bringing light in through new skylights. The length of its longer hall is emphasised by the long, almost mesmeric row of new rafters, and a new courtyard at its end is revealed. New

端部设计了一座新庭院。市场新木门以垂直轴为中心转动,保证空间通风效果,为这座建筑的民俗风格外立面增添了新意。这个项目虽小,但为当地风格吹来了新风。

哥伦比亚格拉莫罗特村遭受了严重的洪水侵袭。在灾后重建工作中,市场重建成为重中之重(166页)。新市场广场的屋顶由混凝土制成;和阿泽尔格河畔拉米尔市场一样,屋顶采用柱子支撑,悬浮于地面之上,为市场提供遮蔽。柱子的设计和几何造型都很优雅,还能收集雨水。屋顶是平的,面积大于市场地面,恰如RPBW在雅典设计的斯塔夫罗斯·尼阿克斯基金会文化中心(2016年)的屋顶一样,呈现出强烈的视觉效果。该建筑另一个与众不同的特点是,市场广场的地面是顺着地势倾斜的。Niro Arquitectura + OAU建筑事务所提出的方案是设计一个实用的功能性楼板,将其划分为多个平台,并在平台间设置蜿蜒的坡道。市场空间是开放的,其中三面有用当地砖砌成的墙体。由于场地的斜坡特征,市场好似剧院般,将人们的注意力引向某一个方向,只不过这个方向是山谷风景,而并非戏剧舞台。该市场的屋顶结构以及对场地的处理堪称杰出之作。

我们对市场的研究始于古希腊,此次调研也直接或间接地对其进行了参考。古希腊广场促进了社会、政治和商业方面的互动。今天,数字化也在向前述乃至其他领域推广,甚至通过线上零售和配送服务构建了食品配送媒介。大众是否希望市场大厅变成网络虚拟空间呢?同时,为供应全球城市型消费社会,城市与乡村之间日益脱节,加剧了工业化对自然环境的破坏。但毋庸置疑的是,谁都应当可以走入市场,从农民那里购买新鲜正宗的当地农产品。著有《站在乡村一边:农业与建筑》(另有同名展览)的法国学者塞巴斯蒂安·马罗特更进一步提出:"如果城乡之间要重新建立一种至关重要的可持续平衡关系,当地市场将是关键,它也是我们打造这一关系的实验室。"

人们应该在意盘中餐食,应该重视城乡联系,应该建设以社区和现实空间相遇为基础的公民社会,这对文明的生存至关重要。为此,我们急需提出相关日程,用心打造市场大厅建筑。

timber doors which open on vertical pivots make the spaces porous and refresh the vernacular look of the facades. This is a modest project that delivers a fresh assertion of local identity.

In Colombia, a replacement market hall was central to rebuilding the town of Gramalote after it was destroyed by floods (p.166). The roof of the new Market Square is a concrete canopy that, as in Lamure-sur-Azergues, is supported by columns and shelters the market that it floats above. The engineering and geometry of the columns is elegant and enables rainwater harvesting. The roof is flat and bigger in area than the building underneath, like RPBW's canopy over the Stavros Niarchos Foundation Cultural Center (2016) in Athens. This creates a powerful visual identity. The other extraordinary feature of the building is that the floor of the market square follows the slope of the site. Niro Arquitectura + OAU found the solution to delivering a practical, functional floor by dividing it into terraces and threading ramps between them. The space is open and walled on three sides with local brick, and the slope makes it like a theater directing attention one way, but to a view of the valley rather than a stage. The building's roof structure and its solution to its site are outstanding.

Our starting point for market places was Ancient Greece, and in our survey, references to it emerge directly and indirectly. The agora fostered interactions that were social, political and commercial. Nowadays, the digital world is advancing into all such interactions and beyond, even becoming a food delivery medium via online retail and delivery services. Do we want market halls to become virtual, another zone in cyberspace? Meanwhile, the growing disconnection between city and countryside drives the industrialised destruction of the natural environment to supply a global urban consumer society. Yet surely, anyone should be able to walk into a market place and buy fresh, authentic local products from a farmer. Sébastien Marot, the French academic and author of "Taking the Country's Side: Agriculture and Architecture" (an exhibition and book), goes as far as to say: "Local markets are key spots and laboratories if we are to reclaim a balanced, vital and sustainable relationship between city and countryside".

We should care about what we eat, value the urban-rural connection and promote a civil society based on communities and encounters in real space. These calls are vital to civilization's survival, and they suggest that market halls need to rise up the architectural agenda.

阿泽尔格河畔拉米尔的有顶市场
The Covered Market of Lamure-sur-Azergues

Elisabeth Polzella Architecte

以当地材料和形式，打造跨水而居的本土先锋派市场
Local materials and form create an avant-garde vernacular in a marketplace built over a river

在法国东部阿泽尔格河畔拉米尔镇的市政厅旁，有一条河流经过那里，将村镇一分为二。当地在跨河大桥上新建了一座有顶市场大厅，取代了曾经的广场。这座有顶市场采用了里昂建筑师伊丽莎白·波尔泽拉的设计，具有坡屋顶，屋顶下是市场区，旁边是藤架或棚道，让人感觉这座市场是村镇与生俱来的一部分。

石与木的对话

与传统有顶市场一样，石头和木材不仅是市场的支撑元素，也构成了市场的框架和屋顶。市场结构采用了当地原材料和本地人熟悉的建筑形式，因而很好地融入了周边环境。维尔瓦西石属于石灰石，产自当地，是世界上最坚硬的石材之一，在这里与玻璃一起构成屋顶。整个市场的框架则采用了当地盛产的花旗松木。市场位于镇中心，是开放的集会场所，也可以举办市政典礼。建筑师沿用地中海市集的说法（"拒绝空谈的艺术"），将这座市场结构描述为"当代拟古主义"或是"本土先锋派"。

结构

主要的有顶空间是一个长方形区域，这里就是市场所在地，上方覆盖坡屋顶，藤架从坡屋顶处一直延伸至市政厅旁。沿长方形两个短边排列的是顶端尖细的石柱，上面是实木梁，它们共同支撑着由檩条组成的水平藤架屋顶和市场的整体框架。框架采用实木材质，按模塑设计形成，并由螺栓连接，檩条和切板上有氯丁二烯橡胶绷带保护层，上面覆盖着扁平的石瓦。坡屋顶处屋顶的上半部分是钢化玻璃瓦。

市场的稳定性可通过柱桥间的连接横向调节。纵向上，部分支柱间设置了高背长椅，加强了支撑能力；另外还有木板橡束框架，进行加固。屋顶南北两侧三角形山墙百叶设计或"耳膜"设计对市场的热舒适性起到了很重要的作用。它们不仅允许热空气对流，还形成了避免北风侵扰的保护层。屋顶南侧部分采用了特定的倾斜度，可以让冬日阳光实现低角度射入，还能抵挡夏日炎热阳光的暴晒。"耳膜"瓦片悬挂在屋顶上，由木架固定，由石质百叶遮盖。藤架屋顶采用木檩条元素，由长石瓦覆盖，少部分与市政厅相连，保证稳定。

有顶市场与藤架为小镇营造了柔和平静的氛围。这要归功于它们的体量与周围环境的完美融合，以及对当地高贵耐用材料的建设性利用。建筑师称之为"恬淡平实、中性风采"。

对这座有顶市场而言，开放结构维持了全景视角，石柱强度保证了经久耐用。通过玻璃瓦进入的光线照亮了顶端中心部分，映射出桥下的流水。继石头和木材之后，自然光成为建筑师采用的第三种材料。

Beside the town hall of Lamure-sur-Azergues in eastern France is a new covered market hall built over the river that divides the village. The site was once a plaza situated on a bridge over the river. This covered market, which comprises a marketplace under a pitched roof and an adjoining pergola or covered walkway, is designed by Lyons-based architect Elisabeth Polzella to give the impression of always having been there.

A dialogue between stone and wood

As with traditional covered markets, wood and stone act as supporting elements, framework and cover at the same time. The market structure is rooted in its location by the origin of its materials and the familiar forms used. Villebois stone, a local limestone that is one of the hardest in the world, is used in conjunction with glass to form the roof cover, while locally-harvested Douglas fir timber frames the whole marketplace.

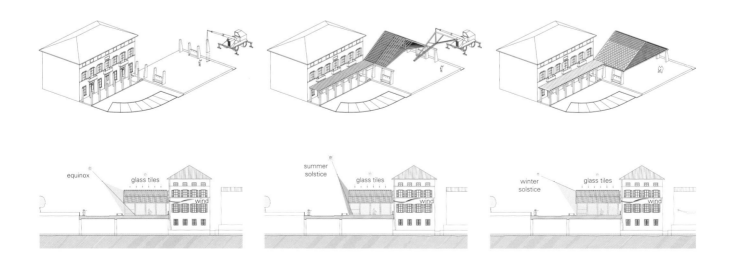

The marketplace is a central, open gathering place that can also host municipal ceremonies. Recalling that rhetoric (the art of speaking effectively) was born in Mediterranean markets, the architects describe the structure as "a contemporary archaism" or "vernacular avant-garde".

Structure

The main covered space is a rectangle hosting the market under a pitched roof, from which the pergola extends beside the town hall. On the opposite short sides of the marketplace are tapering columns or pillars of stone, which carry solid wooden beams pegged upon them. These carry the horizontal roof structure of the purlins for the pergola, and the framework for the market. The frames are solid wood, molded, with bolted connections. The purlins and slashed plates, protected by a neoprene band, carry the cover of flat stone tiles. The upper part of the pitched roof has tempered glass tiles.

The stability of the marketplace is adjusted transversely with links between the pillars and the bridge. Longitudinally, high back benches between some of the pillars help to brace them, while the plates bind the frames. Louvers in the north and south triangular gables or "eardrums" are important for the thermal comfort. They provide protection from the north wind while letting hot air escape by convection. To the south, their inclination lets through low winter sunshine, while protecting from the burning sun in summer. The plates of the eardrums are suspended, stabilized by wooden hangers, and sheltered under the louvers of stone. The pergola roof is covered by long stone plates on timber purlins, and its stability is ensured by occasional links with the town hall.

The covered market and pergola offer a soft and calm presence in the village, thanks to their volumes in harmony with the surroundings, and the constructive expression of the local, noble and durable material. The architect declares it has "no fashion, no style, no gender".

The open structure preserves a panoramic view of the environment, and the stone pillars' strength ensures the durability of the covered market. The glass tiles illuminate the center and the highest part of the place, reflecting the water that flows under it. After stone and wood, light is designed as the third material.

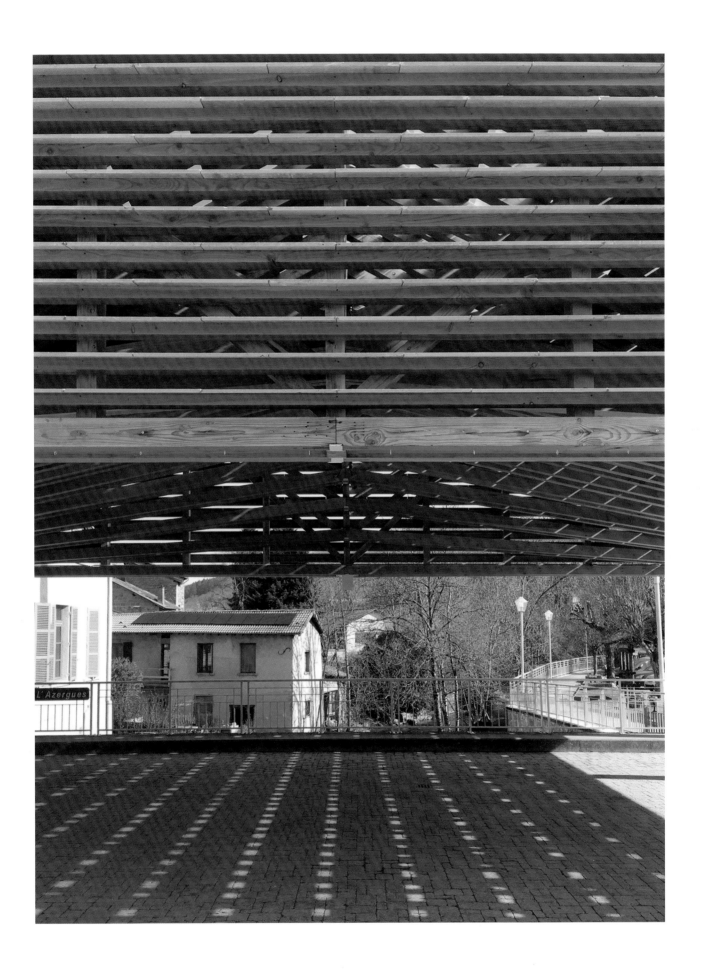

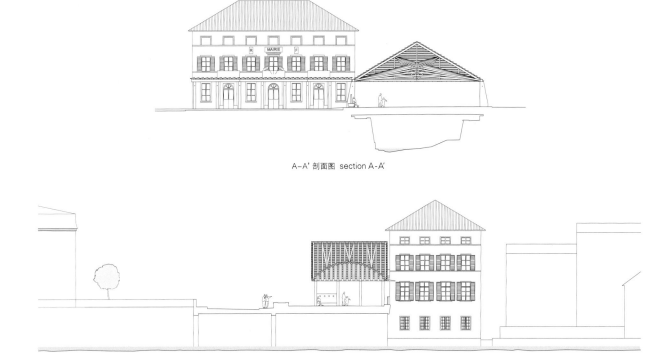

A-A' 剖面图 section A-A'

B-B' 剖面图 section B-B'

一层 ground floor

项目名称：The Covered Market of Lamure-sur-Azergues / 地点：Lamure-sur-Azergues, Rhône department, France / 事务所：Elisabeth Polzella Architecte DPLG / 项目团队：Elisabeth Polzella Architecte, Atelier NAO(architecture & wood structures), GEC Rhône-Alpes(construction economist) / 制造商：Coquaz & Béal(stone structure, glass, zinc), Farjot Toitures(wood framework), MAV'LEC(lighting) / 客户：Lamure-sur-Azergues townhall / 功能：covered hall and pergola annexed to the townhall / 建筑面积：252m² / 材料：villebois stone, douglas timber, glass / 造价：EUR 190,000 / 竞赛时间：2015 / 竣工时间：2017 / 摄影师：©Georges Fessy Photographer (courtesy of the architect) - p.155, p.157[bottom], p.158~159, p.160[upper], p.164[bottom], p.165; ©courtesy of the architect - p.157[top], p.160[lower], p.161, p.163

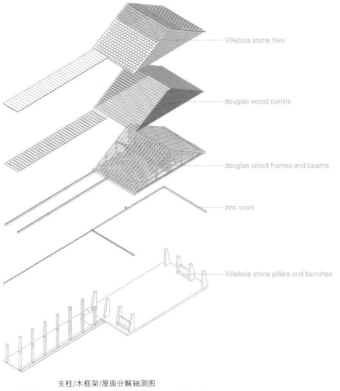

支柱/木框架/屋面分解轴测图
exploded axonometry of pillars / wood framework / roofing

格拉莫罗特市场广场
Gramalote Market Square

Niro Arquitectura + OAU

特色混凝土屋顶市场建于斜坡之上,重建小镇依此凝聚团结
A rebuilt town is brought together by a market built on a steep slope, under a distinctive concrete roof

2010年,一次山体滑坡摧毁了哥伦比亚湿热的内陆小镇格拉莫罗特。然而,人们并没有放弃家园。市政当局对小镇进行了重建,于2017年新建了一座全新的市场大厅,占地900m²,在回归人口和周边农民重整重聚等议题上发挥了重要作用。经过一番竞争后,哥伦比亚首都波哥大的一家建筑事务所成为这座市场的设计方。这家建筑事务所就是现在的Niro Aquitectura+ OAU。

市场大厅采用开放设计,但设有由柱子支撑的屋顶,维护费用低廉,其独特的建筑结构更是承载着小镇居民聚集点的功能。市场由混凝土结构模块组合而成,其中屋顶采用平顶设计,几乎呈正方形外形,由八个矩形模块组成。各模块分为三个部分,各部分均由带柱头的正方形切面柱支撑,柱头垂直边采用弯曲设计,整体形成梯状外形。屋顶进深为20m,支撑柱有24根,隔行交错排列,分为八组,每组三根。屋顶集成了雨水收集装置,同时漏斗形柱子发挥了辅助作用。

市场的用地面积为1300m²,倾斜度至少为1:4,因此柱子的高度不统一,从3m到12.4m不等。屋顶下的市场空间也不完全是同一个平面上的正方形,而是多层级的平台。为避免场地挖掘和垃圾填埋,建筑师设计了沿地形铺设的平台,同时以此沿途提供了不同的视域。为解决交通问题,建筑师在市场上下游出入口之间设计了连续的之字形小路。这条小路由一系列平行坡道组成,坡度为1:10;坡道与各平台的梯台一起发挥作用,方便市场内的手推车和轮椅通行。

市场广场下方有一个灵活的多功能平台,农贸市集、美食广场、市集和音乐活动等都可以在那里举办。它还可以用作一个观赏周围景观和洛斯阿皮奥斯峡谷的观景台。市场设有70个商业摊位,按所售货品分为两类。

市场大厅三面采用黏土砖外立面,不接触屋顶,并与之形成鲜明对比。墙体上设有部分穿孔。市场高处的外立面始于整体结构中唯一的弯角。顾客穿过外立面后就来到了市场的最高层。墙面上端延伸至屋顶之外,下端向斜坡花园开放。台阶逐级向下,直至开放平台。黏土砖是在附近的库塔小镇制造的,使立面凸显本地特色。

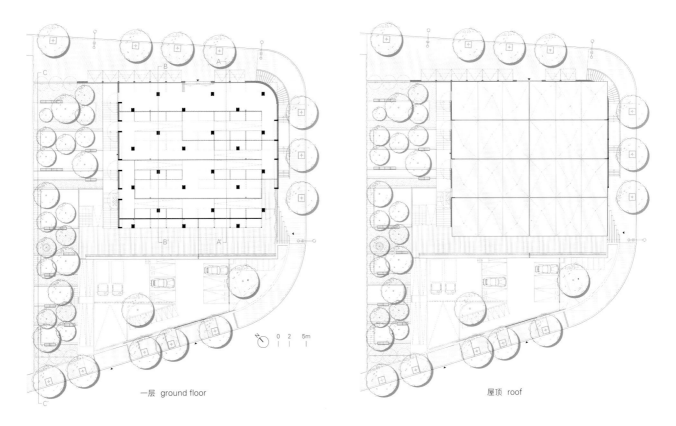

一层 ground floor 屋顶 roof

项目名称：New Gramalote Market Plaza / 地点：Gramalote, Colombia / 事务所：NIRO arquitectura + OAU / 设计建筑师：Jheny Nieto Ropero, Rodrigo Chain
设计团队：Fabrizzio Milano / 客户：Fondo de Adaptacion / 设计监理：Sociedad Colombiana de Arquitectos / 施工方：Arquitectura y Concreto
施工主管：Jose Herrera / 技术空间：Rodrigo Montaño / 顾问：Proyectar ingenieria / 总楼面面积：1,300m² / 竣工时间：2017
摄影师：©Sergio Gomez (courtesy of the architect)

The small town of Gramalote in the humid interior of Colombia was destroyed by landslide in 2010, but it was not abandoned. The municipality has rebuilt, and a new 900m² market hall completed in 2017 has been important in re-integrating the returning population and nearby farmers. Its design followed a competition won by architects based in the capital Bogotá, who are now Niro Aquitectura + OAU.

The low-maintenance market hall is open air but covered by a canopy roof supported by columns. This distinctive architectural structure signals a meeting point for the town's inhabitants. It is made up of a concrete structural module, in which the flat, almost-square roof is formed by eight rectangular modules. These have three sections, and each is supported by a square-plan column with capitals in which the vertical sides curve out into a trapezoid at the roof. The 24 columns are arranged in eight lines of three across the 20m-deep roof, but alternate rows are staggered. A rainwater collection mechanism is incorporated into the roof and facilitated by the funnel-like columns.

The 1,300m² site slopes with a gradient of at least 1:4, so the columns vary in height, from 3m to 12.4m. The market space beneath the roof is not a flat square but multi-leveled in terrace platforms. To avoid intervention with excavations and landfills, platforms that respect the terrain are incorporated,

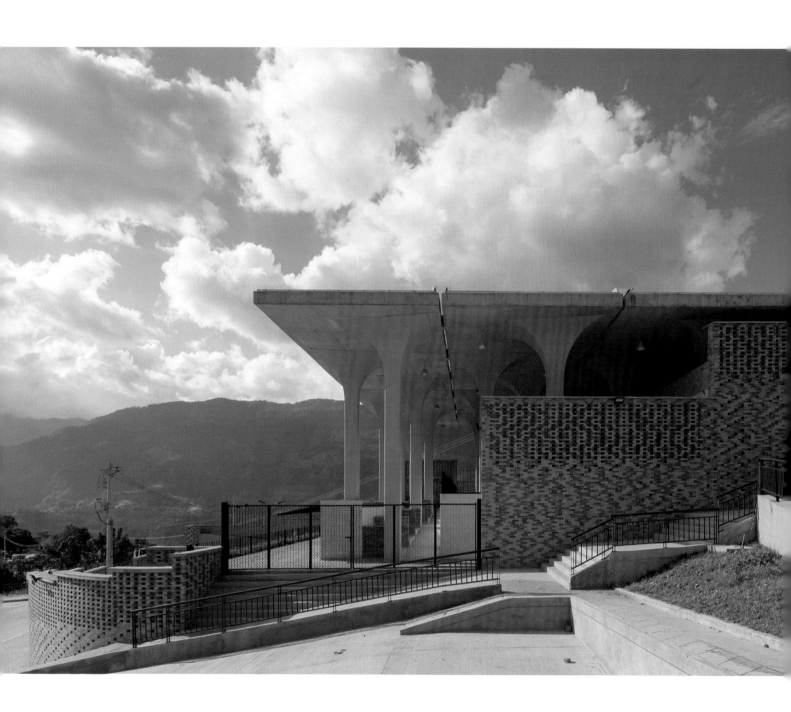

and offer different views along the route. The solution to circulation was to create a continuous zigzag path between the higher and lower access points of the market. This path is a series of parallel ramps with slopes of 1:10, which along with landings on the platforms, enable the passage of a market cart or wheelchair.

The lower part of the market square has a flexible multi-use platform in which a farmer's market, a food court, bazaars, musical events, etc. can operate. It also functions as a viewing platform looking towards the landscape and the Los Apios gorge. The market has 70 commercial stalls, which comprise two types depending on the products sold.

The market hall has clay brick facades on three sides which contrast with the roof and make no contact with it. Parts of the walls are perforated. The facade at the upper part of the site starts with the only curved corner in the structure. Access through it leads to the highest market hall level, and the wall's other end extends beyond the roof. On this side, the downward facade walls open onto sloping gardens and steps descending to the open platform. Clay blocks were manufactured in the nearby town of Cucutá, which helps the facade to generate it a sense of local identity.

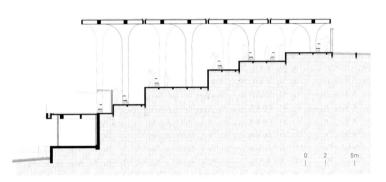

A-A' 剖面图 section A-A'

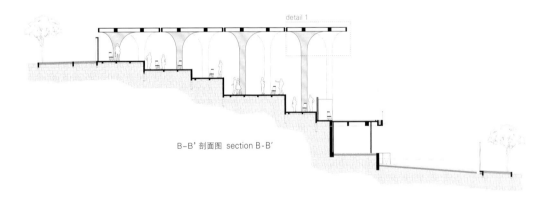

B-B' 剖面图 section B-B'

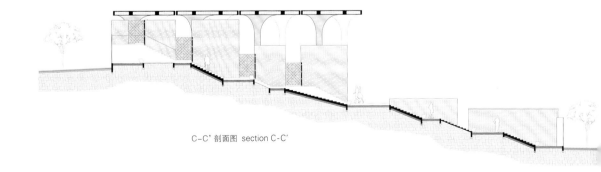

C-C' 剖面图 section C-C'

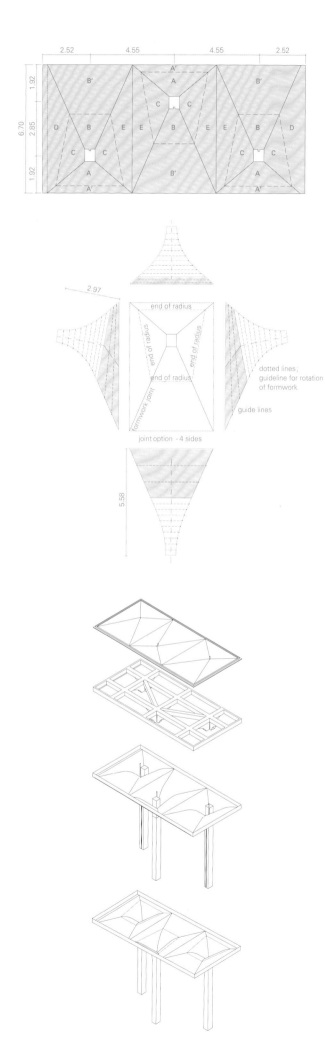
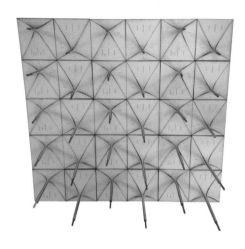

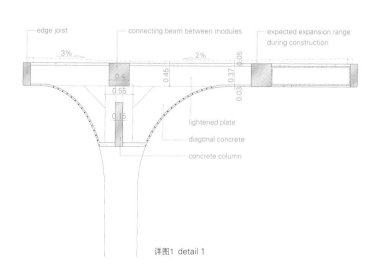

详图1 detail 1

圣雷莫安诺纳里奥市场重建
Restructuring of the Sanremo Annonario Market

Calvi Ceschia Viganò Architetti Associati

利用木框摊位快速恢复市场活力,重建设计带来现代意大利市场大厅设计新风
Rapid renewal based on wooden frame trading units refreshes a modernist Italian market hall

安诺纳里奥市场位于意大利里维埃拉区圣雷莫镇镇中心,曾是繁华的食品市场。但近年来,受制于缺乏维护、供应商补货基建不足和停车问题,该市场日渐衰退。市场大楼为钢筋混凝土建筑,面积达3300m²,采用现代主义风格,由Cesare Fera和Luciano Grossi Bianchi设计,于1960年完工,其特点在于中央市场大厅上方为抛物线拱形屋顶和天窗。

按要求,当地建筑事务所Calvi Ceschia Viganò要在四个月内恢复安诺纳里奥市场。由于设计咨询与协作人员不足,缺少捍卫私人利益的"委员会",因此在当地开展咨询工作要面临不少问题,于是,事务所决定换种方法,让当地已然疏远的市场—消费者关系重新热络起来。他们得出的结论是,城市市场必须尽力让自己的吸引点与众不同,因为消费者光顾市场的目的不仅是省钱,还要寻求产品最佳性价比。

市场重建项目的第一阶段是实施紧急维护,进行功能干预和设备工程干预。后续阶段包括重组道路系统、将工会办公室搬到楼上以创造更多市场空间,并重新铺设市场外行人空间。

接下来的工作仅耗时3个月,第二阶段之后最明显的变化体现在市场大厅中。食品仍由市场边缘区的商店出售,这一点保持不变;而亟需整改的是市场大厅,那里是主要活动场所,水果和蔬菜货摊不断延伸占地,杂乱无章。商贩过去拥有宽阔的开放柜台,其中有很多还聚在霓虹灯柱下面。现在,商贩的营业单元采用了开放式木框架,并设在略微高于地面的平台上,平台表面由树脂砖重新铺设而成。市场一共有18个这样的标准单元,它们在大厅中分成三排,这样的设计加强了内部的交通效率,为大厅带来了秩序。在另一个较大的中央单元中,设计师在周边空闲区布置了一个咖啡吧台,这里提供休闲桌、长凳和垃圾箱。

经过划分后,每个标准单元可容纳四个商贩,每个商贩都有自己的自来水、标牌、照明和消防系统。交易时间结束后,下拉遮帘,即可关闭交易单元。市场中设置了几种色调柔和的遮帘,每个摊位使用的都是其中的一种颜色。整个市场大厅如同整洁有趣的包厢,给人留下了深刻的印象。鉴于木材经济耐用、灵活方便,建筑师为营业单元和大门入口选用了木材,便于定制操作,体现了老市场振作革新的理念,让顾客舒心放心,犹如大家购买的蔬菜水果一般,有机天然。新建单元隔断了个体交易者的聚集,让顾客能够清晰方便地选择在哪里买、买什么产品。

The Mercato Annonario was once a bustling food market at the heart of the Italian Riviera town of Sanremo, but in recent years the market has fallen into decline due to lack of maintenance, the lack of infrastructure for vendors to renew stock, and parking problems. The 3,300m² reinforced concrete modernist building was completed in 1960, designed by Cesare Fera and Luciano Grossi Bianchi. It features a central market hall under a parabolic vaulted roof and clerestory windows. Local architects Calvi Ceschia Viganò was given the brief of reviving the market within a four month timeframe. Local consultation was challenging due to the lack of a community of people able to sit at the design table, and "committees" defending private interests. The architects decided on an approach to regenerate the local relationships that had failed. They concluded that city markets must try to differentiate their offer as much as possible, since the approach of market customers is not only to save money but also to look for the best possible relationship between quality and price of the products.

将建筑改造成一个专营食品的经济文化中心。
transformation of the building into an economic and cultural center dedicated to food

1 Loft和"LANTERNA"改造
  · 销售产品且现场交货
  · 市场扩建
  · "La Lanterna"是一个新的城市空间，可进行完全与食品有关的展示与活动

2 室外区域布局
  · 定义域市场相连区域的铺地材料
  · 新植被

3 重新布局每周开放的市场
  · 市场扩建
  · 商贩停车位

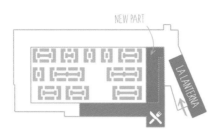

1 MODIFICATION OF THE LOFT + "LANTERNA"
- sale of products and offer on site
- market expansion
- "La Lanterna" as a new civic space dedicated to demonstrations and events entirely dedicated to food

2 ARRANGEMENT OF EXTERNAL AREAS
- definition of the pavement connecting the spaces
- new vegetation
- new image and characterization of the place

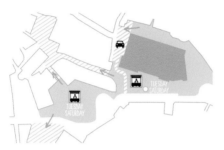

3 REORGANIZATION OF THE WEEKLY MARKET
- market expansion
- parking spaces reserved for traders

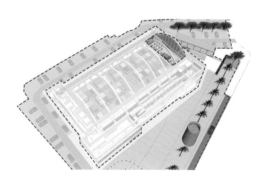

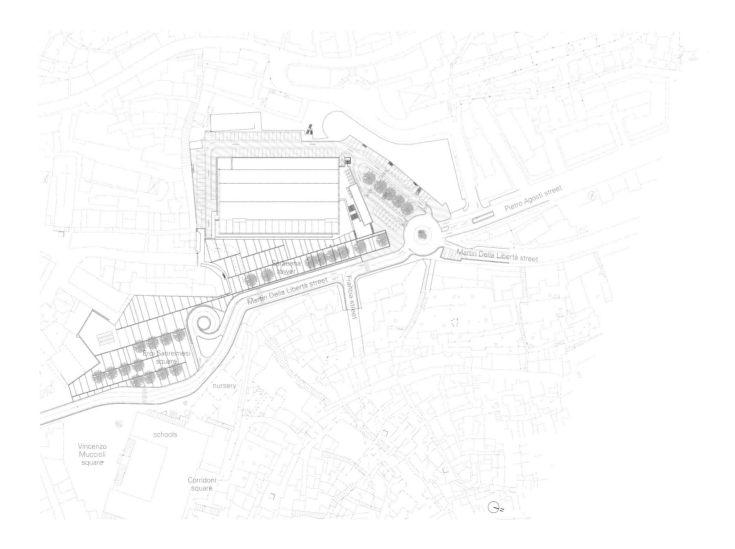

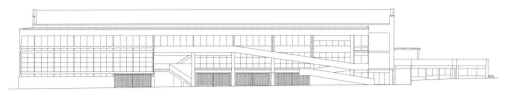

东立面 east elevation

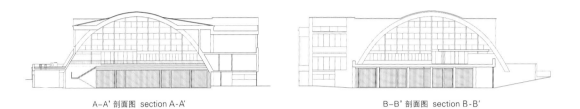

A-A' 剖面图 section A-A'   B-B' 剖面图 section B-B'

C-C' 剖面图 section C-C'

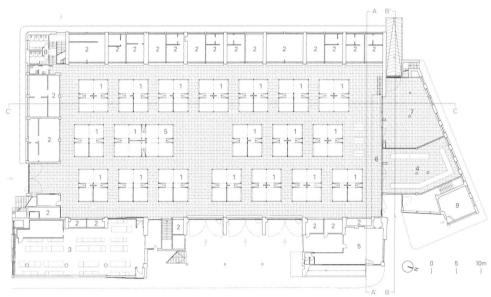

1. 市场摊位 2. 市场箱形空间 3. 海鲜店铺 4. 当地农产品摊位 5. 吧台 6. 书籍交流区 7. 废品收集站 8. 卫生间 9. 兽医诊室
1. market stalls 2. market box 3. fish shop 4. local farmer 5. bar 6. book crossing 7. waste collection 8. toilets 9. veterinary

一层 ground floor

项目名称：Restructuring of the Sanremo Annonary Market / 地点：Sanremo, Italy / 事务所：Calvi Ceschia Viganò architetti associati
当地建筑师：Marco Calvi, Gaetano Ceschia, Andrea Viganò / 结构工程师：Siccardi Luca / 电气工程师：Paolo Magna
合作者：Sara Marino, Cinzia Loiacono / 承包商：Giò Costruzioni srl / 客户：Comune di Sanremo / 总楼面面积：3,300m²
竣工时间：2019 / 摄影师：©Aldo Amoretti (courtesy of the architect)

The first phase of the project was the implementation of urgent maintenance, functional and plant engineering interventions. Further phases include reorganization of the road system, moving trade union offices upstairs to create more market space, and repaving external pedestrian spaces. Following works that took just three months, the clearest changes after phase two are in the market hall. Food was sold from shop units around the periphery, which remain unchanged, but the main activity was on the market hall floor, where the continuous spread of fruit and vegetable felt disorganised. Vendors had wide open counters, many clustered around posts supporting neon lights. Vendors now operate from open wooden frame units that are on platforms slightly raised from the floor, which has been resurfaced using resin tiles. There are 18 such standard units, separated and arranged in three rows across the hall, in a configuration that improves circulation and brings order to the market hall. An additional larger central unit includes a café counter serving tables in the adjacent clear space, where benches and bins are also provided.

Each standard unit is partitioned to house four market traders, each with their own running water supply, as well as signage, lighting and fire-fighting systems. After trading hours, each unit closes its trading side by drawing down a blind, which is in one of the several pastel colors. This creates the impression that the market hall is a field of neat, playful boxes. The choice of wood for units and also for the entrance doors was made for its economy, durability and flexibility. It allows for customization and communicates the idea of renewal. It is reassuring to customers, creating associations with "organic" and all that is natural, like the fruit and vegetables that many come to buy. Separating the individual traders in the new units makes their choice of where and what to buy clear and convenient.

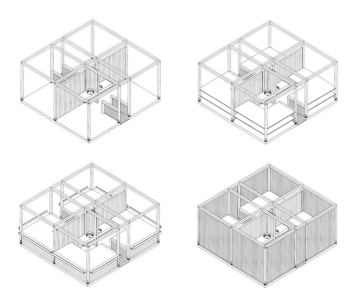

胜利市场临时安置点
Temporary Site of Shengli Market

LUO Studio

以低成本的玻璃结构，打造中国通风良好的模块化临时市场大厅
A low-cost glazed structure creates an airy temporary modular market hall in China

胜利市场为中国北方内陆城市濮阳的当地社区服务了几十年，变得又脏又乱，造成了交通堵塞。市场重构将涉及拆除、改造和重建，这不仅需要时间，还会扰乱市场，无法充分满足人们必不可少的基本生活需求。对此，解决方案是建立一个临时市场来替代旧市场。

幸运的是，穿过繁华的胜利东路，就在旧市场的对面，有一块闲置土地可供短期出租。它非常靠近旧市场，便于商贩腾挪和市民购物。该地块从胜利东路处后退，预留出一个停车区，用地平面呈方形。由于新市场大厅占地为2902m²，远小于旧市场，所以新建筑平面布局采用占满场地原则。

临时建筑与可持续利用
客户要求快速施工、成本低廉，建筑师则非常重视可持续性，希望该建筑完成临时安置点的使命后，依然可以进行功能的转换和再回收利用。建筑师决定采用平屋顶配全玻璃箱式结构。该结构采用标准化部件搭建，外形如工业化蔬菜种植棚。所有部件轻量化、模块化、预制，可快速建造，通过低成本而获得相对大的空间，其中的杆件是标准件，可以被拆除、回收、再利用。

从无序到有序
大多数常规菜市场的问题就是视觉标识混乱。如何通过必要的增建，在无序的空间里营造出秩序，方便贩卖管理和市民采购，是一项重要的议题。一方面，新结构必须要尊重和利用现有结构和构造杆件的模数；另一方面，新结构采用了非标准部件，这种部件再利用性弱，因此造价必须尽可能低。

市场空间加入了封闭式商店和开放式货架/柜台，并在边缘区域安排了多个相连的商店。封闭式店铺选择三面围合设置，每个铺面为4m×4m的开间和进深，与外墙的结构杆尺寸相协调。这些商店规模较大，一店一名，主要销售干货、熟食和调味料。此外，每家店铺的门都是卷帘门，门的上檐口有凸出卷轴，由角钢和木格栅遮蔽防护，上面安装了尺寸统一的标识招牌，而且招牌的安装高度完全相同。

货架区域是开放的，商户得到的平均空间有限。每个货架单元长度为2m，一个商户一般可占用两个货架。由于从视觉高度和长度上，都不可能就着货架做标示，因此设计师结合市场大厅的结构柱，设置了伞状倒置正四面锥体钢木架，既可安装市场不同分类的标牌，同时方便安装灯具照明。

市场还增加了入口雨篷。为了确保其稳定性，立面上安装了伞状倒置正四面锥体钢木架结构，不但提供了支撑，同时也节省了材料。

此外，在市场区的后面还建有一排鱼用水族箱和一个厕所区。

所有建材均选取低价而易得的材料，包括普通木材、轻型钢板、水泥板、角钢、聚碳酸酯板，这些材料既方便安装，也利于建造。暖色的天然木材在整个空间中随处可见，比如，开放的货架、封闭店铺的檐口、伞状的结构柱，这样的设计增强了空间的辨识度，营造出场所的秩序感。

Shengli Market has served a local community in the inland city of Puyang in northern China for decades, but the original market became dirty and messy, and it created traffic jams. Rebuilding would involve demolition, transformation and re-construction, which takes time, disrupting a market that meets people's indispensable basic living needs. The solution was to build a temporary market to substitute for the old one. Happily, an unused plot of land was available for short-term rental, situated opposite the original market across the busy Shengli East Road, making it convenient for vendors to move and citizens to shop. The plot is set back from the road, leaving a parking area and forming a square plan. Since the new 2,902m² market hall is far smaller than the old market, the new construction fully occupies the site.

Temporary architecture & sustainability
The client required a rapid construction process and low cost, while the architects attached great importance to sustainability, hoping the architecture could be used for other purposes or recycled and reused after it completes its role as a temporary market. The architects decided to adopt a flat-roofed fully glazed box structure like an industrialized vegetable-growing shed, built with standardized component sets. All the components are lightweight, modular and prefabricated, which ensured rapid construction, reduced cost and resulted in a relatively expansive architectural space. Those construction elements including standardized rods can be dismantled, recycled, or reused for other constructions.

From disorder to order
Most traditional food markets have a common problem – chaotic visual identity. Within the architectural structure, it was important to create order in the disorderly space by adding some necessary "extensions", to facilitate the management of selling activities and improve citizen's shopping experiences. On the one hand, these extensions must respect and take advantage of the standard modular sizes of existing structures as well as rods. On the other hand, built with non-standard components (less reusable), the extensions' cost must be minimized. Enclosed shops and open shelf/counter trading are inserted into the space. Multiple connected shops are arranged on the edge areas of the space. Each shop is enclosed on three sides, and has an equal width and length of 4 meters, which is coordinated with the sizes of the external walls' structural rods. Those shops have a relatively large scale and independent names, mainly selling dried food, cooked food and season-

Shengli east road

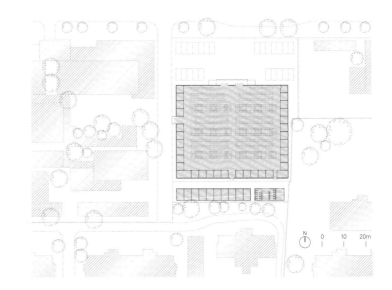

ings. Besides, every shop has a rolling door, and on the door's top is a protruded shaft, which is covered and protected by steel angles and wooden grilles, on which shop signboards of uniform size are installed at the same height.

The shelf trading area is open, and the average space for each trader is relatively limited. Each single shelf has a total length of 2 meters, and generally one trader occupies two shelves. Considering the visual height and length, signage indicating what is sold by the traders is unfeasible on the shelves. Therefore, based on the structural columns in the open space, steel and timber are used to create overhead inverted rectangular pyramid structures, which resemble umbrellas. Those structures not only serve for signage to indicate the market's different areas, but also are convenient for lighting fixture installation.

An entrance canopy is also added. In order to ensure its steadiness, rectangular pyramid frame structures cantilevering from the facade provide support, which are material-saving as well. In addition, a row of units for aquatic tanks for fish and a bathroom block are built behind the market block.

All the extensions are built with cheap and easily accessible materials including ordinary timber, lightweight steel panels, cement slabs, steel angles and polycarbonate sheets, which are easy for installation and construction. Natural and warm-colored timber can be widely seen within the entire space, which is applied to the open shelves, cornices of the enclosed shops and umbrella-shaped structural columns, thereby resulting in a clear visual identity of the space and creating a sense of order in it.

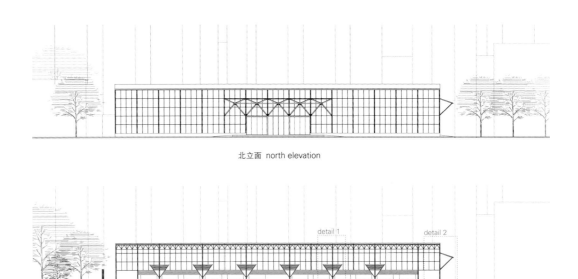

北立面 north elevation

A-A' 剖面图 section A-A'

1. 开放货架
2. 封闭货架
3. 停车空间
4. 水产品区
5. 女士卫生间
6. 男士卫生间
7. 垃圾收集站

1. open shelf
2. closed shelf
3. parking space
4. aquatic housing
5. ladies' toilet
6. gentleman's toilet
7. garbage-station

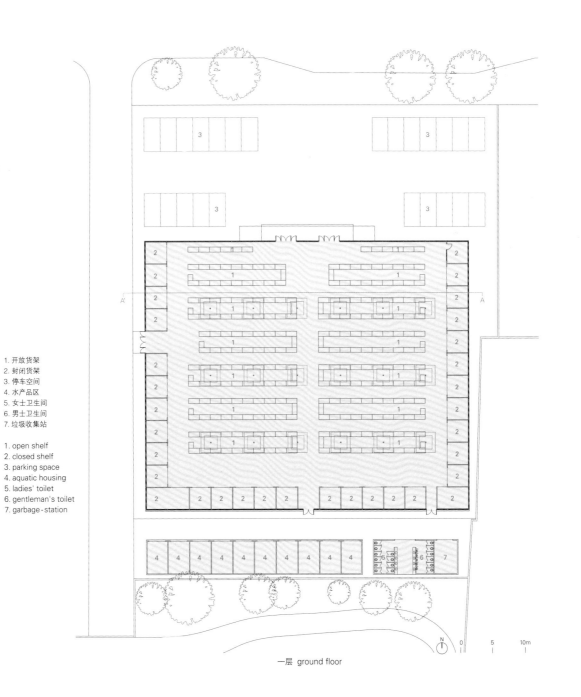

一层 ground floor

项目名称：Temporary Site of Shengli Market
地点：Southeast of the intersection of Shengli Road and Changqing Road, Puyang City, Henan Province, China
事务所：LUO studio / 主创建筑师：Luo Yujie
参与设计师：Wei Wenjing
协调设计公司：Shanghai QIWU Architectural Design & Consultation Co., Ltd.
建造公司：Puyang JINGYI Architectural Decoration Design and Engineering Co., Ltd.
客户：Shengli Sub-district Office, Hualong District, Puyang, Henan, China
总楼面面积：2,902m² / 建造时间：2018.12—2019.4
摄影师：©Jin Weiqi (courtesy of the architect)

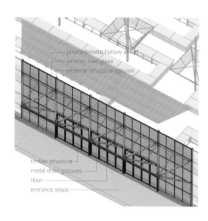

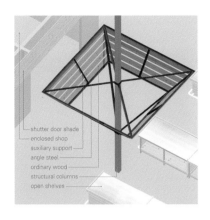
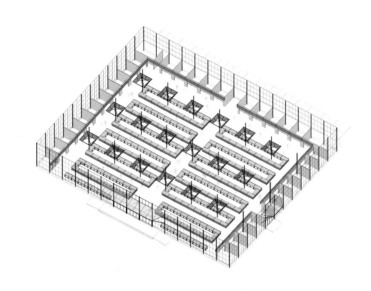

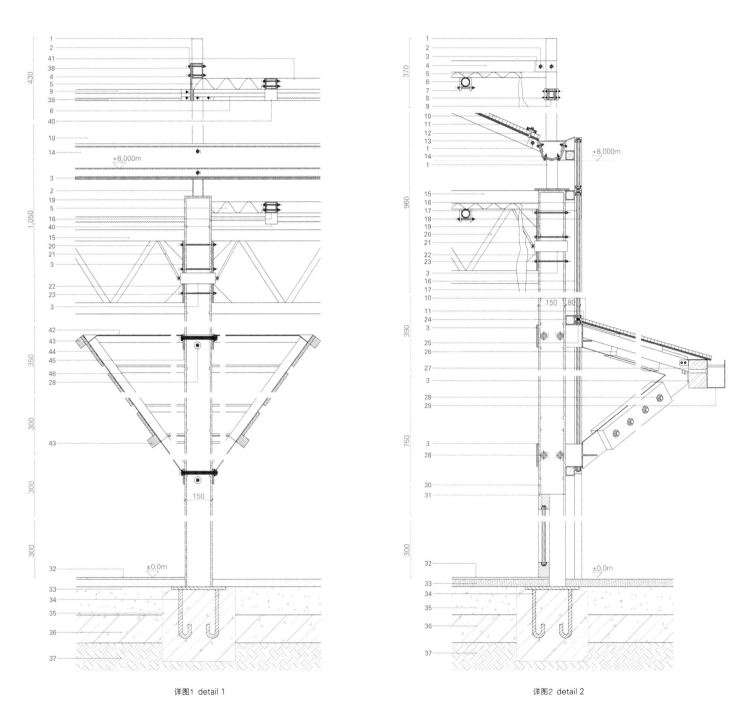

详图1 detail 1 　　　　　　　详图2 detail 2

1. 60mm x 60mm x 2mm hot-dip galvanized square pipe 2. M10 bolt 3. connecting member of 10mm thick galvanized hot plate 4. 60mm x 80mm x 2mm hot-dip galvanized square pipe 5. sunshade cloth 6. push rod 7. push rod support 8. M8 bolt 9. 50mm x 30mm x 2mm hot-dip galvanized square pipe 10. 6m x 2.1m x 20mm sunshine board 11. the rubber gasket 12. tapping screw 13. aluminum layering 14. water chute 15. 70mm x 50mm x 2mm hot-dip galvanized square pipe 16. 50mm x 50mm x 1.5mm hot-dip galvanized square pipe 17. double layer hollow tempered glass 6mm x 9mm x 6mm 18. aluminum alloy partition 19. 150mm x 150mm x 3.5mm hot-dip galvanized square pipe 20. angle rib 21. Phi 10mm hot-dip galvanized steel 22. 10mm thick hot galvanized sheet 23. M16 bolt 24. 50mm x 50mm x 2mm hot-dip galvanized square pipe 25. contains desiccant spacers 26. 120mm x 120mm anti-corrosion wood square 27. 100mm x 150mm anti-corrosion wood square 28. M20 bolt 29. 2mm hot galvanized plate guide tank 30. structural sealant 31. the metal frame 32. 10mm thick 600mm x 600mm floor tile 33. 40mm thick cement mortar binding layer 34. Phi 16mm rebar 35. reinforced concrete embedded parts 36. 150mm thick reinforced concrete foundation 37. element of soil compaction 38. bearing pedestal 39. rack 40. gear 41. drive line 42. national standard angle steel 50mm x 50mm x 3mm 43. LED lights bring 44. 135mm x 15mm x 3660mm anticorrosive board 45. 10mm thick diagonal fixture 46. auxiliary support

纳吉克罗斯市场大厅
Market Hall of Nagykőrös

Kiss-Járomi Architect Studio

市场大厅半透明立面内嵌字母结构，凸显市场功能
Letter-shaped structures set in a translucent facade of a market hall spell out its function

　　纳吉克罗斯是一个小镇，也是匈牙利最后一批家畜市场的所在地之一，却没有市场大厅。对此，布达佩斯建筑师久洛·吉斯和伊恩·杰罗米从古希腊城市生活聚集区——广场汲取灵感，兴建新的市场建筑。不过，与广场不同的是，这座总占地面积达935m²的市场大厅并非开放空间，而是当之无愧的当代封闭空间。在这座小镇中，新市场大厅可谓是非比寻常，这不仅是因为镇上的500座建筑几乎都是保存完好的老建筑，还要归功于它风格迥异的外立面。

　　按设计理念，这座市场大厅是一座拥有长长的玻璃外立面的建筑，同时前后两个玻璃立面外都有第二层立面，这层立面为复合铝板材质，采用圆形穿孔设计。前立面嵌入四个彼此分离的凸出的结构，结构的一层是入口空间，设有内门、外门，上面还有一个小小的二层空间。这四个结构被设计成字母外形，为"P I A C"，即匈牙利语"市场"。其中，只有字母"A"向上延伸，突破了双层外立面的高度。

　　晚间，大厅内的LED灯可通过外立面照亮周边；日间，周边的自然光又能通过外立面照亮大厅内部，外立面的圆形穿孔区能在一天中的不同时间，在大厅内外投射有趣的光影图案。

　　在超长坡屋顶的相交处，隆起了三角体屋脊，而天窗就沿此排列。日光通过这里，洒向市场大厅内部。屋顶另装有光伏板。天窗可以打开，这有利于水果和蔬菜所需的充分通风。

　　颜色在建筑设计中起到了核心作用。外立面采用了特殊的"玛雅金"装饰效果，而字母形入口、屋顶和天窗结构表面则采用了青铜色表面；两者凸显了黄金质感，散发着金属光泽，给人以温暖的感觉，同时对光线也非常敏感。

场地总面积为2659m²，市场大厅后面有一个开放的区域，这里的一棵老栗子树被保留下来。开放区域可容纳一个农贸市场，可通过市场大厅后立面上三个齐平设置的入口进入，后立面在高度上只有一层。市场大厅一侧的尽头还保留了一座老建筑，作为辅助区使用，这里提供安保和公共卫生间等功能。

建筑师总结说："我们想为这座城市建造一座独一无二的建筑，它要充满新意，引人注目，令人喜爱，吸引城市中年轻人，还要起到新城市活动中心的作用。零能耗和独一无二的穿孔复合立面设计是建筑的亮点。"

Nagykőrös is a small town that hosts one of Hungary's last livestock markets, yet it had no market hall. The inspiration for Budapest-based architects Gyula Kiss and Irén Járomi for a new market building came from the ancient Greek agora, which was a gathering place where city life was focused. Unlike an agora, the market hall is not an open space but a distinctly contemporary enclosed volume of 935m². It is unusual not only because the town's 500 buildings are almost all old and protected, but also because of its distinctive facade.

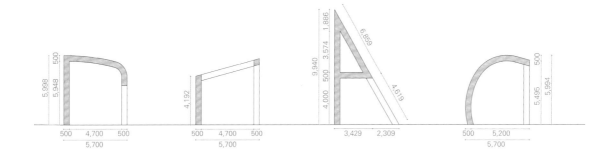

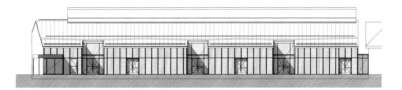

北立面 north elevation

西立面 west elevation

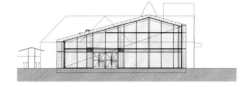

东立面 east elevation

南立面 south elevation

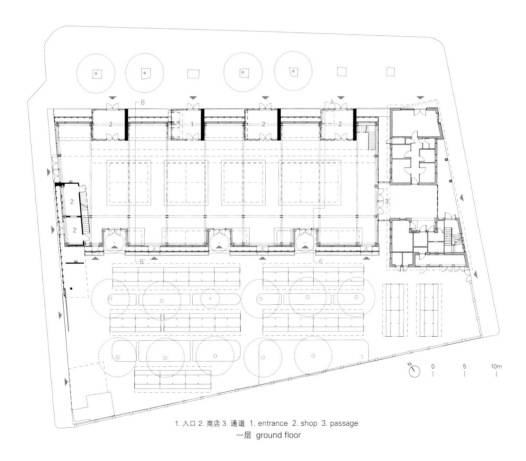

1. 入口 2. 商店 3. 通道 1. entrance 2. shop 3. passage
一层 ground floor

项目名称：Market Hall of Nagykőrös / 地点：Nagykőrös, Hungary / 事务所：Kiss-Járomi Architect Studio / 设计：Irén Járomi, Gyula Kiss / 员工：Ádám Tillinger / 支撑结构：Salacz Ákos – TervS LTD / 建筑工程：Lovas Albert – Optiterv Ltd / 建筑电气：Csaba Bakos – Bakos Villamosmérnöki Tervező Iroda LTD / 弱电：Bulla Péter Alfréd / 消防：Attila Kiszely / 总承包商：EU-Építő Ltd / 钢结构：TESZ-97 LTD / 玻璃墙：SHÜCO / LTD-BAU LTD / 复合板与屋顶：Prefa Hungária LTD / 客户：The Government Of Nagykörös City – Szabolcs Czira Mayor / 用地面积：935m² / 总楼面面积：2,659m² / 竣工时间：2019 / 摄影师：©Croce & WIR (courtesy of the architect) - p.198~199, p.202, p.203, p.205bottom, p.207; ©György Palkó (courtesy of the architect) - p.196~197, p.200; courtesy of the architect - p.205top

The concept is a long building with a glass facade, and a second outer facade on the front and rear sides, made of aluminum composite panels perforated with circles. Four structures penetrate the front facade and are spaced out across it, containing entrance spaces with inner and outer doors, and small second floor spaces. They are shaped as letters to spell "P I A C", the Hungarian for market. Only the "A" rises above the double facade.

The facade allows the entire volume to shine outwards with its internal LED lighting in the evening, and brings natural light into it by day. The field of perforated circles in the outer facade creates a playful pattern of light and shadow outside or inside, depending on time of day.

The market hall also receives daylight through skylights made by windows along a triangular-sectioned spine structure rising above the ridge where the long pitched roofs meet. Photovoltaics are carried by the roof. The skylight windows are openable, contributing to sufficient ventilation for the fruits and vegetables.

Color plays a central role. A special "mayagold" finish to the facade, and bronze in the letter-shaped entrances, roof and skylight structure, both produce the effect of gold – metallic, warm and responsive to light.

The total site area of 2,659m² includes an open area behind the market hall, where old chestnut trees have been saved. The area can host a farmer's market, and is accessed by three doors set flush in the rear facade, which is single story in height. The market has also retains an old building at one end of the hall, which now hosts market support functions such as security and public toilets.

The architects conclude that "we wanted to give the city a unique building that is fresh and eye-catching, yet lovable and captivates the youth of the city. We have added a new center for new urban activities. This is served by the zero energy design and the one-of-a-kind perforated composite facade design".

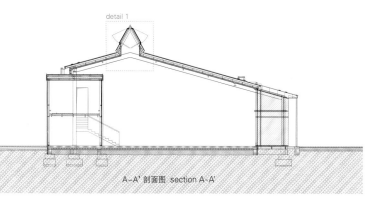

A-A' 剖面图 section A-A'

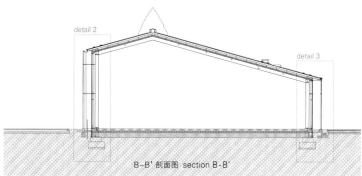

B-B' 剖面图 section B-B'

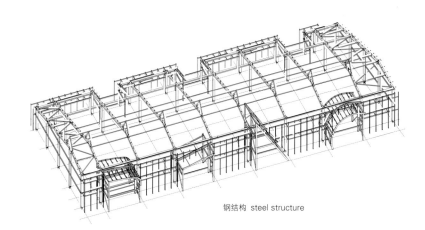

钢结构 steel structure

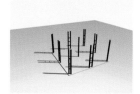

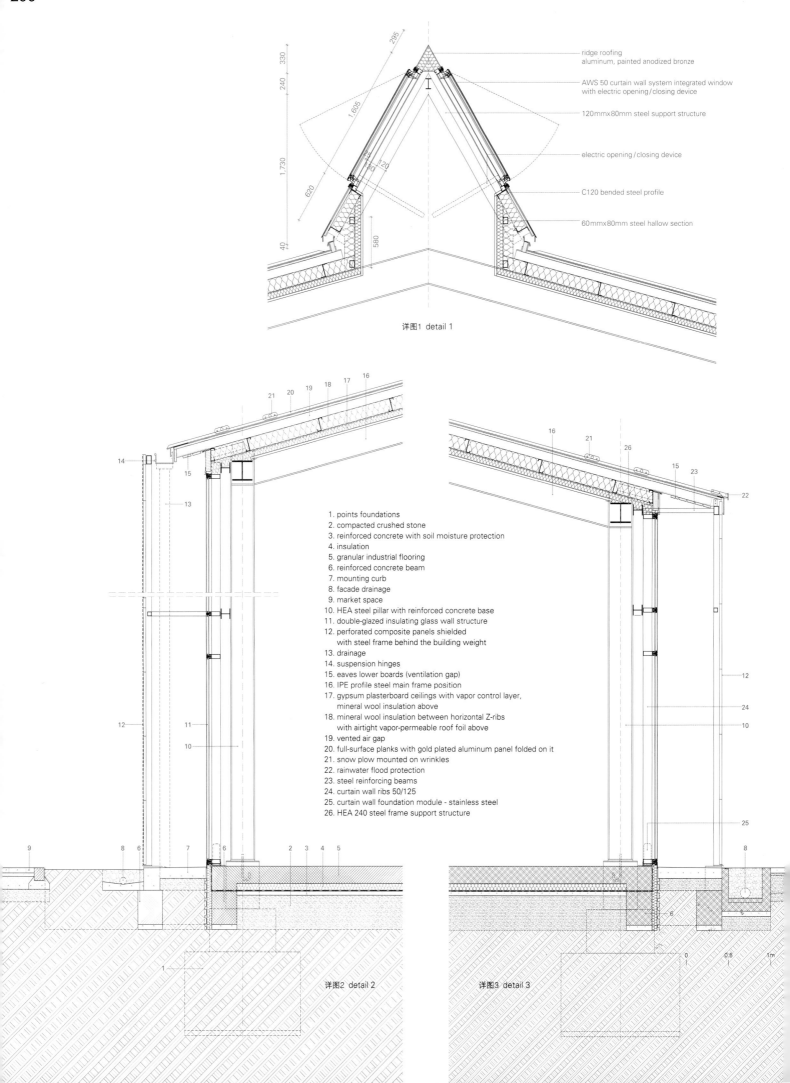

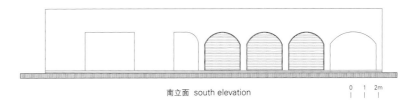
南立面 south elevation 0 1 2m

特拉斯科工匠市场
Tlaxco Artisans Market

Vrtical

自然光线和天然木材使废弃的市场大厅焕然一新，创造了全新的工匠市场
Natural light and wood refresh a disused market hall to create a new artisan market

在墨西哥小镇特拉克斯科被正式定为促进旅游业的"魔法小镇"后，市政府决定建立一个手工艺品市场来吸引游客。然而，当时的预算不足以建造一座全新的建筑，因此市政府决定翻新20世纪80年代建成的旧集市。旧集市的主要问题是采光不好和空气不流通，因此，墨西哥城的Vrtical建筑事务所决定只复原原有的外墙和基础，保留主体外壳，翻新墙壁，添加新的饰面，并打造一个全新的屋顶。

建筑内部分为两个主厅，二者呈L形布局。大主厅里有14个摊位，由来自市区各处的工匠们经营，并通过木架子展示产品。小主厅里有一个容纳多个工作间的空间，还有一个可灵活使用的活动空间。两个区域都通向一座新的庭院。这不仅是采光和通风解决方案的一部分，也创造了供各种活动使用的乐园。从原木大厅的纯玻璃端望去，人们可以看到一座干旱作物花园。

从正外立面进入市场，我们将走过一道柱廊，穿过一扇扇圆拱门和一扇矩形门，体味当代风格与当地风情的碰撞交融。其中，部分大门是原本就有的，还有部分则是新建的，主要目的是突出新旧风格的有趣对话，并让建筑尽可能地通风透气。针对圆拱门设计的木门则与矩形旋转门一样，都是按中轴旋转开门，方便消费者进入小主厅。柱廊位于玻璃屋顶下，是令人愉悦的会面地点。

两个主厅的上方都建造了承重墙和松木屋顶桁架。室内可以享受自然光的照射，这得益于与主厅同长的天窗。后者不仅增强了照射进市场的光线，也衬托出了让当地居民感到自豪的标志性建筑。线性天窗的设计灵感来自由路易·卡恩设计的德州沃思堡金伯利博物馆的天窗。

市场保留了乡土气息，同时又在阳光和开放设计的衬托下焕然一新。椽子、门和市场摊位使用的木材，使整个空间材质一致，与市场上出售的工艺品相互呼应。

After the small Mexican town of Tlaxco was officially designated a "Magic Town" to promote tourism, municipal authorities decided to build an artisan market attractive to visitors. The budget was insufficient for a completely new building, so it was decided to recycle the old market built in the 1980s. Its main problem was the lack of light and ventilation, therefore the architects, Mexico City-based Vrtical, decided to recover only the pre-existing outer walls and foundations, to keep the main shell, renovate the walls, add new finishes and create a whole new roof.

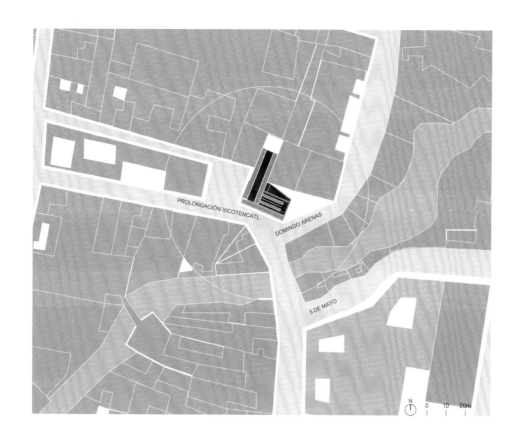

Internally, the building is organized into two main halls which form an L shape. The larger one houses fourteen stalls, for artisans from various parts of the municipal district. The stalls have timber shelves to present products. The smaller hall has a space for workshops, and offers a flexible space for events. Each area is connected to a new courtyard. This is a part of the solution to the lack of light and ventilation, and also creates interesting spaces to bring activities. A garden planted with arid species is visible through the fully glazed end of the log hall. A colonnade walkway across the front facade has a series of round arches and a rectangular frame that creates a dialogue between contemporary and vernacular languages. Some of these come from the original project and the architects also create new ones to make a playful dialogue with the old and the new, and to make the border of the building as porous as possible. Wooden doors shaped for the round arches are centrally pivoted to open into the smaller hall, as do pivoted rectangular doors. The colonnade walkway is now sheltered under a glass roof, and becomes an attractive meeting point. Load-bearing walls and laminated pine roof trusses are built over both halls. The interiors enjoy natural lighting thanks to skylights along the length of the hall that magnify the sunlight throughout the market place, at the same time as creating an iconic structure which the local population can feel proud of. The linear skylights are inspired by those of the Kimbel Museum in Fort Worth, Texas, by Louis Khan.

The market retains a vernacular feel, but is refreshed with light and openness. The use of wood in rafters, doors and market stalls introduces a consistent materiality throughout the space, and resonates with the artisan offerings of the market.

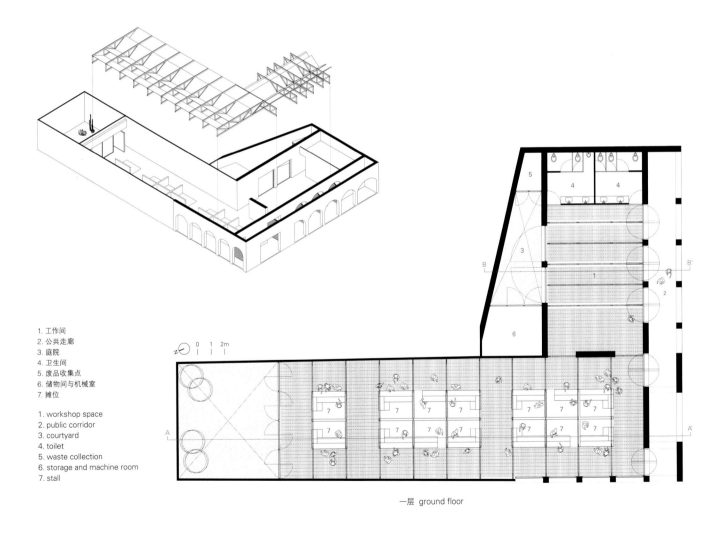

1. 工作间
2. 公共走廊
3. 庭院
4. 卫生间
5. 废品收集点
6. 储物间与机械室
7. 摊位

1. workshop space
2. public corridor
3. courtyard
4. toilet
5. waste collection
6. storage and machine room
7. stall

一层 ground floor

项目名称：Tlaxco Artisans Market / 地点：Tlaxco, Mexico / 事务所：Vrtical / 设计者：Luis Beltrán del Río, Andrew Sosa / 施工：Jorge Rivera / 合作者：Alejandro Solano, Dolores Galicia, Hector Rojas, Ana Lourdes Esquer, Ximena Rebollo, Arelly Blas / 用地面积：534m² / 总楼面面积：410m² / 竣工时间：2017 / 摄影师：©Rafael Gamo (courtesy of the architect) - p.209top, p.210~211, p.213, p.216~217, p.218~219; ©Enrique Márquez Abella (courtesy of the architect) - p.208~209bottom p.212, p.214

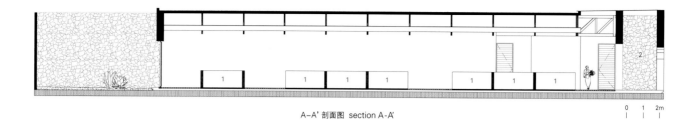

A-A' 剖面图 section A-A'

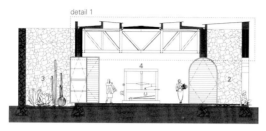

1. 摊位
2. 公共走廊
3. 庭院
4. 工作间

1. stall
2. public corridor
3. courtyard
4. workshop space

B-B' 剖面图 section B-B'

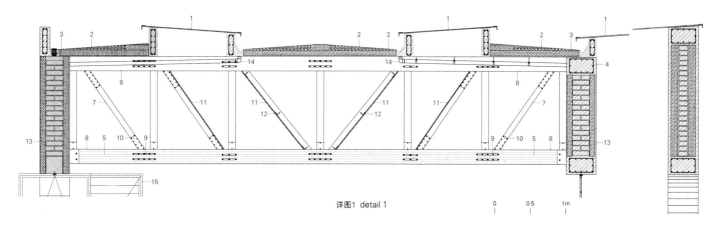

详图1 detail 1

1. 9mm glass with safety film placed over 25mm x 25mm and 6mm thickness steel clip angles, supported over Norton tape, sealed against steel clip angles with synthetic rubber sealant 2. wear course of gravel Tezontle stone with final layer of compacted earth (tepetate) and asphalt saturated felt, 4% slope 3. 102mm prefabricated expanded polystyrene panel 4. concrete beams with reinforced bars 5. glue-laminated finger joint wood truss 6. 150mm x 75mm top chord of double glue-laminated finger joint system pine wood beam 7. 75mm x 75mm diagonal web member of finger joint system pine wood beam 8. 150mm x 75mm bottom chord of double beam finger joint system pine wood 9. Trought bolts 10. 50mm x 50mm steel clip angles attached to beam by flat head wood screw, automotive paint final finish phosphor color 11. parasol made by drywall of 13mm sealed and white paint finish, 13mm gypsum board for bouncing light, white paint final finish 12. 25mm x 25mm steel clip angles frame to support parasol 13. 102mm Covitec panel roof 240mm x 120mm x 60mm red brick wall core 102mm prefabricated expanded polystyrene panel to thicken wall 14. stormwater drain pipe 15. pivot door 16. gauge metal plate, mounted on a 2x4" PTR frame, protected with primer and finished with automotive paint

Durbach Block Jaggers Architects

Is managed by the three directors Neil Durbach[right], Camilla Block[center] and David Jaggers[left]. As co directors for over twenty five years, they work together seamlessly across all stages of their projects.
Has many years working with existing buildings, sensitive or highly visible contexts, heritage or otherwise, in a respectful and poetic manner. Aims for an architecture that is both confident and courteous. Environmentally sustainable design principles permeate the logic of durable design in siting, landscaping, management of services and detail design. Received the Robin Boyd Award three times for the best residential project in Australia, as well as State and National Architecture Awards across all building types, including public, education, commercial, residential, heritage and urban design.

©Daniel Boud

gad

As an atypical design studio born in the transformation of rapid development of Chinese architecture design revolution, gad is one of the most influential design companies countrywide. Instead of mass production, they would rather choose to slow down and search for the existing significance of every space. Cooperate with insightful proprietors from diverse fields like hotel design, housing design, theater design and so on to boost new ideas for architecture, which at the same time, encourages itself to step into a better design world.

RANDJA - Farid Azib Architects

Farid Azib received a master's degree at the Architectural Association School of Architecture in London. Set up the studio Randja in Paris, 2007, after ten years' experience as Project Leader at Alain Derbesse Architectes. Acquired an in-depth experience both with public and private facilities including cultural center, school, leisure center, water sports center, airport terminal, office space, and lodging. Has made his efforts to complete project managements that were often difficult and sometimes conventional. His creative approach is based on constraints, which are the driving forces behind a more enjoyable and vibrant architecture.

Ljubomir Jankovic

Is Professor of Advanced Building Design at the University of Hertfordshire, UK. Conducts interdisciplinary research into advances of building performance design and urban systems design, combining the fields of engineering, geography, computer science and complexity science, and challenging existing ways of thinking. Established Zero Carbon Lab at the University as a unique interdisciplinary research group where design thinking and advanced computer modelling methods are brought together. Obtained his Dipl.Ing./MSc from the University of Belgrade, and a PhD from the University of Birmingham, both in Mechanical Engineering.

Junya.Ishigami+Associates

Junya Ishigami (1974) studied architecture at Tokyo University of the Arts. Began his career at SANAA (2000-2004) before founding his studio, junya.ishigami+associates, in Tokyo in 2004. The firm gained international recognition following the completion of the Kanagawa Institute of Technology's KAIT Workshop in 2007. Has been awarded the Golden Lion for Best Project at the 12th International Architecture Biennale in Venice. Has recently seen constant growth in foreign attention for a series of striking exhibitions and architecture projects, such as 'Freeing Architecture' at the Foundation Cartier pour l'Art Contemporain in Paris, France, and the Art Biotop Water Garden, Tochigi, Japan. Its projects are approached by a limitless and wide creative perspective capable to render and deliver unique and inspiring outcomes.

CHROFI

Was founded in 2000 as Choi Ropiha following the win in the international competition for the re-design of the TKTS booth in Times Square, New York in 2008. The practice in Manly is home to a team of 26 professionals headed by directors John Choi, Tai Ropiha and Steven Fighera. Other projects of note are Lune de Sang, Stamford on Macquarie, Manly Masterplan and Binhai New City International Competition (1st Prize). Publications include A+T Strategy:Public (Spain), C3 Energy Efficiency - a New Aesthetic (Korea), journals and newspapers.

THAD

Founded in 1958, the Architectural Design and Research Institute of Tsinghua University (THAD) is a national Grade A architectural design institute. Engages in a wide range of services, including engineering design for public and civil buildings, urban design, residential planning and housing design, urban master planning, heritage conservation and rehabilitation, landscape design, interior design, early-stage feasibility study and architectural programming, as well as engineering consulting, catering to national and social needs in all areas and on all fronts.

Phil Roberts

Is a design writer based in Montreal, Canada. Also works as a design consultant for various companies in the creative industries. Has an Honours Bachelor of Arts from the University of Toronto, where he majored in architectural design and minored in Canadian studies and Spanish.

Vrtical

Was founded in 2014 by Luis Beltran Del Rio[left] (1982) and Andrew Sosa[right] (1992) in Mexico City. They seek to satisfy the needs of the clients by understanding various contexts and scales and fine-tuning the processes and methods for creating spaces. Luis Beltran graduated from the National Autonomous University of Mexico. Received a master's degree in urban management for developing countries from the Technical University of Berlin. Andrew Sosa graduated from the Universidad Anáhuac México. Both strive to create efficient, functional, sustainable and beautiful designs.

Calvi Ceschia Viganò Architetti Associati

Founded by Marco Calvi[center], Gaetano Ceschia[left] and Andrea Viganò[right], with the aim of networking professional skills in various fields. Their goal is to create a flexible structure to activate ad hoc work units efficiently responding to market needs. They create customized structures according to needs and manage them professionally. The team is made up of competent professionals with expertise and flexibility.

MAD Architects

Founded by Chinese architect Ma Yansong in 2004, is a global architecture firm committed to developing futuristic, organic, technologically advanced designs that embody a contemporary interpretation of the Eastern affinity for nature. With a vision for the city of the future based in the spiritual and emotional needs of residents – MAD endeavors to create a balance between humanity, the city, and the environment. Ma was named one of the "10 Most Creative People in Architecture" by Fast Company in 2009, and selected as a "Young Global Leader (YGL)" by World Economic Forum(Davos Forum) in 2014. MAD's signature cultural projects include Harbin Opera House (2015), China Philharmonic Concert Hall (under construction), Ordos Museum (2011). Has on-going international projects located, respectively, in Rome, Paris, Japan, Los Angeles and Beverly Hills. Ma graduated from the Beijing Institute of Civil Engineering and Architecture, and holds a Master's Degree in Architecture from Yale University. Is currently a professor in Beijing University of Civil Engineering and Architecture.

Niro Arquitectura

Jheny Nieto graduated from Javeriana University in 2002, Bogotá, Colombia. From the beginning of her career, she was involved in public space and public building design that had a significant impact on the transformation of Colombian cities. After gaining experience as an architect and project leader in several architectural offices in Bogotá and Medellin, she moved to London in 2008 and earned a master's degree in architecture with the Colfuturo Scholarship. In 2012, she founded NiroArquitectura where she worked in the public and private sector. Has eight years of teaching experience at Javeriana University.

John Wardle ©Pier Carthew

OAU

Rodrigo Chain Rodriguez, co-founder of OAU | Office of Architecture & Urbanism, graduated form Javeriana University in Bogota, Colombia and the Architectural Association School of Architecture in London, UK. Gained experience in a number of high-profile practices such as Heatherwick Studio and Lorenzo Castro Arquitectos. Has lead the completion of many public and private projects including Granada Sur Day Center, Gramalote Market Plaza, Gran Bazar Public Market, e57 Apartments, and amongst others.

John Wardle Architects

Is internationally renowned for making extraordinary buildings and places that matter. The team of 80 design professionals work across Australia and internationally from our two studios in Melbourne and Sydney. Is a large collaborative environment where every project has a range of creative, technical and strategic contributions from a diverse team of architects and interior designers within JWA. Many projects by JWA have been awarded. Has twice received Robin Boyd Awards for best residential project in Australia in 2012 and 2013. In 2018, this included National AIA Awards for Educational Architecture and Interior Architecture and the RIBA Award for International Excellence. In 2020, John Wardle was awarded the Gold Medal, the Australian Institute of Architects' highest honour.

P186 LUO Studio

Luo Yujie is the founder of LUO studio, who also teaches the course of Construction Basics in the School of Architecture at Central Academy of Fine Arts (CAFA). He is committed to creating more durable, friendly and quality spaces with creative thinking, a spirit of craftsmanship and the principle of caring for nature. His works have been shortlisted and awarded by numerous design award competitions worldwide, including but not limited to: WAN AWARDS 2019 - Wood in Architecture (Gold Winner), ABB LEAF AWARDS 2019 - Best Achievement in Environmental Performance (Winner), AR Emerging Architecture Awards 2019 - Shortlisted, A+ AWARDS 2019 - Concepts - Plus-Architecture +Technology (Finalist), and Dezeen's Top 10 Chinese Architecture projects of 2018.

Xaveer De Geyter ©Mirjam Devriendt

XDGA

Is a Brussels-based office practising architecture, urbanism and landscape design founded in 1988 by Xaveer De Geyter(1957) after his experience as project architect for OMA/Rem Koolhaas. During the thirty years of its existence, XDGA has managed to build up a significant portfolio and obtain worldwide recognition thanks to its unique approach, diversified expertise andinternational team (54 collaborators from 11 countries). XDGA counts to this day 5 monographs, numerous awards (Mies Van der Rohe Award, Bigmat Award, Flemish Culture Award for Architecture) and 3 travelling solo exhibitions.

Herbert Wright

Writes about architecture, urbanism and art. Graduated in Physics and Astrophysics from the University of London. Is contributing editor of UK architecture magazine Blueprint, and has contributed to publications including Wired, RIBA Journal, The Guardian, l'Architecture d'Aujourdhui, Abitare and C3. His books include London High (2006) and Instant Cities (2008). Curated Lisbon's first city-wide public architecture weekend Open House 2012, was short-listed to curate Oslo Architecture Triennial 2016, was a 2017 Graduate Thesis juror at SCI-Arc, and keynote speaker at Element Urban Talks, Krakow 2018.

Elisabeth Polzella

P154 Elisabeth Polzella Architecte

Born in 1977 in Saint-Martin d'Hères, Isère, France, Elisabeth Polzella studied at the Architecture School of Grenoble, France and the Univesity La Sapienza of Rome. Worked as project manager of Gilles Perraudin from 2000 to 2003. Since 2012, she has established her own office and continued to work mainly on architectural projects based on history and places using local resources such as stone, wood and earth. Has been teaching at the Architecture School of Lyon, France since 2015, Université des sciences et IUT de Corte, France from 2010 to 2015, the Montpellier School of Architecture since 2003, and the Grenoble School of Architecture since 1996.

Kiss-Járomi Architect Studio

Was founded by Irén Járomi and Gyula Kiss in Budapest, 1990. Both members of the same graduating class at University, and worked together on the renovation project of the Budapest opera house. Irén Járomi always knew that she wanted to be an architect. She loved entering drawing contests as a child, winning several. For young Gyula Kiss, things were different: When he was 16, he stood in the local library and a book fell off the shelf. When it landed at his feet, it was open to a picture of Frank Lloyd Wright's Fallingwater house. With the belief that it is the architects' job to find the answers that guide us into the future and true architects are like conductors, they work together in the whole process of researching, experimenting, making and submitting proposals.

墙体设计
ISBN: 978-7-5611-6353-5
定价: 150.00 元

新公共空间与私人住宅
ISBN: 978-7-5611-6354-2
定价: 150.00 元

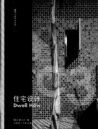
住宅设计
ISBN: 978-7-5611-6352-8
定价: 150.00 元

老年住宅
ISBN: 978-7-5611-6569-0
定价: 150.00 元

小型建筑
ISBN: 978-7-5611-6579-9
定价: 150.00 元

文博建筑
ISBN: 978-7-5611-6568-3
定价: 150.00 元

流动的世界：日本住宅空间设计
ISBN: 978-7-5611-6621-5
定价: 200.00 元

创意运动设施
ISBN: 978-7-5611-6636-9
定价: 180.00 元

墙体与外立面
ISBN: 978-7-5611-6641-3
定价: 180.00 元

空间与场所之间
ISBN: 978-7-5611-6650-5
定价: 180.00 元

文化与公共建筑
ISBN: 978-7-5611-6746-5
定价: 160.00 元

城市扩建的四种手法
ISBN: 978-7-5611-6776-2
定价: 180.00 元

复杂性与装饰风格的回归
ISBN: 978-7-5611-6828-8
定价: 180.00 元

企业形象的建筑表达
ISBN: 978-7-5611-6829-5
定价: 180.00 元

图书馆的变迁
ISBN: 978-7-5611-6905-6
定价: 180.00 元

亲地建筑
ISBN: 978-7-5611-6924-7
定价: 180.00 元

旧厂房的空间蜕变
ISBN: 978-7-5611-7093-9
定价: 180.00 元

混凝土语言
ISBN: 978-7-5611-7136-3
定价: 228.00 元

建筑入景
ISBN: 978-7-5611-7306-0
定价: 228.00 元

新医疗建筑
ISBN: 978-7-5611-7328-2
定价: 228.00 元

内在丰富性建筑
ISBN: 978-7-5611-7444-9
定价: 228.00 元

建筑谱系传承
ISBN: 978-7-5611-7461-6
定价: 228.00 元

伴绿而生的建筑
ISBN: 978-7-5611-7548-4
定价: 228.00 元

大地的皱折
ISBN: 978-7-5611-7649-8
定价: 228.00 元

在城市中转换
ISBN: 978-7-5611-7737-2
定价: 228.00 元

锚固与飞翔——挑出的住居
ISBN: 978-7-5611-7759-4
定价: 228.00 元

创造性加建：我的学校，我的城市
ISBN: 978-7-5611-7848-5
定价: 228.00 元

文化设施：设计三法
ISBN: 978-7-5611-7893-5
定价: 228.00 元

终结的建筑
ISBN: 978-7-5611-8032-7
定价: 228.00 元

博物馆的变迁
ISBN: 978-7-5611-8226-0
定价: 228.00 元

微工作·微空间
ISBN: 978-7-5611-8255-0
定价: 228.00 元

居住的流变
ISBN: 978-7-5611-8328-1
定价: 228.00 元

本土现代化
ISBN: 978-7-5611-8380-9
定价: 228.00 元

气候与环境
ISBN: 978-7-5611-8501-8
定价: 228.00 元

能源与绿色
ISBN: 978-7-5611-8911-5
定价: 228.00 元

体验与感受：艺术画廊与剧院
ISBN: 978-7-5611-8914-6
定价: 228.00 元

记忆的住居
ISBN: 978-7-5611-9027-2
定价: 228.00 元

场地、美学和纪念性建筑
ISBN: 978-7-5611-9095-1
定价: 228.00 元

殡仪类建筑：在返璞和升华之间
ISBN: 978-7-5611-9110-1
定价: 228.00 元

苏醒的儿童空间
ISBN: 978-7-5611-9182-8
定价: 228.00 元

都市与社区
ISBN: 978-7-5611-9365-5
定价: 228.00 元

木建筑再生
ISBN: 978-7-5611-9366-2
定价: 228.00 元

© 2021大连理工大学出版社

版权所有·侵权必究

图书在版编目(CIP)数据

社区市场建筑 / 丹麦BIG建筑事务所等编；于风军等译. — 大连：大连理工大学出版社，2021.5
 ISBN 978-7-5685-2979-2

Ⅰ．①社… Ⅱ．①丹… ②于… Ⅲ．①社区－建筑设计 Ⅳ．①TU984.12

中国版本图书馆CIP数据核字(2021)第068074号

出版发行：大连理工大学出版社
　　　　　（地址：大连市软件园路80号　邮编：116023）
印　　刷：上海锦良印刷厂有限公司
幅面尺寸：225mm×300mm
印　　张：14.25
出版时间：2021年5月第1版
印刷时间：2021年5月第1次印刷
统　　筹：房　磊
责任编辑：杨　丹
封面设计：王志峰
责任校对：张昕焱
书　　号：978-7-5685-2979-2
定　　价：298.00元

发　行：0411-84708842
传　真：0411-84701466
E-mail：12282980@qq.com
URL：http://dutp.dlut.edu.cn

本书如有印装质量问题，请与我社发行部联系更换。